# Pulse Attack

## The Real Story
## Behind the Secret Weapon
## that Can Destroy North America

### Anthony Furey

Published by 2016 Magna Carta
All rights reserved by the author.
ISBN: 978-0-9939195-3-4

To my sons, Andrew and Christopher,
that they may only know peace in their time.

ANTHONY FUREY

*Attack him where he is unprepared*
-      *Sun Tzu, The Art of War*

# SHOTS FIRED IN THE DARK

Picture children playing in the park after school. People driving home from work, turning up the volume on their favourite music. At home others are watching television or doing the laundry while dinner cooks in the oven. It's just a regular day for the many millions of people living in the United States and Canada.

This is the sort of day the attack will happen. The day life in North America as we know it comes to an end. When nobody expects it. And there will be no bombs raining down. No soldiers in the streets. It'll have come out of the blue.

Although there have been warnings, hints, that this was to come. We just haven't paid much attention to them. We'll wish we'd done more as we look back at recent events with the benefit of hindsight. Like what happened one evening in California several years ago.

Coyote Valley is a largely undeveloped tract of California land on the southern outskirts of San Jose. The 101 freeway, connecting Los Angeles to San Francisco, cuts right through it. Nearby are lush county parks and 18-hole golf courses. It's a beautiful area.

Passersby probably never give a second thought to the unassuming collection of grey electrical transformers that sit off to the side of the highway. They'd no doubt find it hard to believe that at that spot on April 16, 2013, an event occurred that continues to puzzle and worry some of North America's leading national security experts to this day.

It was a quiet night at the Pacific Gas & Electric Company's Metcalf transmission substation. Even though it's one of the largest electric utility companies in America and its combined operations provide electricity to over 5 million customers, there were no security guards posted at the facility that night. Just like on most nights. There weren't any other employees on shift either.

The only eyes on the ground were the surveillance cameras and motion detectors and someone clearly knew how to avoid their detection because just before 1 a.m. a person was able to sneak in and cut the telecommunications cables leading into the station. The phones went dead.

Thirty minutes later, a group of people opened fire on the station with military-grade rifles. They continued firing for 19 minutes – targeting the transformers. When they finished firing their weapons they didn't enter the facility. They didn't steal anything. They just left as mysteriously as they had arrived.

While the shooting was going on, people slowly discovered something was wrong. Nearby residents found their internet was down – one of the casualties of the cut cable. A motion detector at the plant went off, likely sensing the ricochet of bullet shells. This sent an alert to someone at the company. At the same time one of the transformers, its exterior damaged by bullets, started leaking oil. An equipment failure notice was automatically sent to a company office almost a hundred miles away. Finally, a technician at a different facility down the road heard the gunfire and called police.

Officers arrived only one minute after the last bullet was fired. But the attackers were already gone. The officers didn't see them leave or even hear any of the blasts. All they had to work with was the 911 call about shots fired. So, seeing nothing suspicious, they shrugged it off and left.

A couple hours later – soon after 3 a.m. – an employee finally arrived to check things out. What he found was doubtless a surprise. While there are a number of different types of equipment failures the electricity industry is used to dealing with, this wasn't one of them.

The employee discovered in those early morning hours that a total of 17 transformers had been riddled with gunfire. Over 50,000 gallons of oil had leaked out around the station. Power had to be rerouted to avoid a blackout in the area. The most troubling evidence was the more than one hundred shell casings found littering the ground.

Who were these unidentified figures, systematically firing their weapons into the dark of the California night, targeting the metallic monoliths that powered nearby homes and businesses? Were they disgruntled former employees giving the finger to an old boss? Petty criminals looking to steal anything they could resell? One of California's notorious gangs or drug rings? Operatives of a foreign terrorist group, prepared to die for a chance at taking down the Evil Empire?

It could have been any of the above. Or none of the above. To this day, nobody knows the answer to the mystery. And that's not for lack of trying.

The Federal Bureau of Investigation looked into it. They told media they didn't believe it was an act of terrorism. In a press release, PG&E stated their facility had been hit by "vandals". But mere vandals typically don't carry military grade AK-47 assault rifles – which is the type of weapon that carries the shell casings that were found on the ground.

Other industry and government officials also probed the incident and came up with little evidence. "While we have not yet identified the shooter, there's some indication it was an insider," a Department of Homeland Security rep told CNN in 2015. Even two years after the incident, that was and still is the best guess available.

The attackers left no trail behind them. There was zero evidence aside from the shell casings. Zero hint as to their motivation or goals. And this unsolved mystery has left a small but informed number of academics, politicians, industry experts and intelligence operatives deeply concerned for the safety of their country.

But why? Just how big of a problem could this attack have caused? Even if they'd blown up the transformer station entirely, and even if they were terrorists, how bad would the effects from this one incident have been?

Surely the police would eventually stop the attack with limited casualties. And surely a blackout to that region would be solved in a matter of days, if not hours. The electrical companies would just re-route power to that part of California from other areas. It would be little more than a mild inconvenience. The economic impact would be slim to nil. In terms of scope, compared to all the other threats North America faces, the Metcalf transformer station at first glance seems like a small affair.

At first. The physical impacts of this one attack aren't what keep experts worried about the incident to this day. It's what this represents, what it might have been foreshadowing and paving the way for, that keeps people up at night. That's because they know the San Jose gunmen weren't the only malicious presence at work that night. A foe was watching from above while America slept.

Kwangmyŏngsŏng is the Korean word for Bright Star. It's an inviting sounding phrase. The sort of name someone might give to christen his first sail boat. The international community didn't feel like bringing out champagne, however, when the dictatorship of North Korea launched the satellite Kwangmyŏngsŏng 3-2 into orbit on December 12, 2012.

It was the rogue state's first ever successful satellite launch. Earlier incarnations of what's referred to as the KSM series failed to reach orbit. The previous one blew up merely 90 seconds after launch earlier that year, on April 13. The Democratic People's Republic of Korea said the KSMs were nothing but observation satellites designed for weather forecasting purposes. But the rest of the world wasn't buying it.

Governments around the world – including the United States and Canada – denounced the launch. They saw it as a thinly veiled attempt at creating a ballistic weapons program to launch strikes against other countries. The United Nations Security Council had previously passed sanctions against North Korea for its nuclear weapons testing. And after the KSM 3-2 went up they passed resolution 2087 against the launch to add to the laundry list of condemnations the country faced.

If it was just a glorified weather balloon, why were hundreds of thousands of soldiers dancing in the streets of Pyongyang after it finally made it to orbit? Why did North Korea put so much fanfare behind it? The original April launch date they'd planned was actually the 100th anniversary of founding dictator Kim-Il Sung's birth. Then there's the fact the name Bright Star is no random phrase. It represents the myth that when the second dictator Kim Jong-Il was born a "bright star" shone in the sky to herald his arrival.

North Korea is a country notorious for hyping up the three generations of their dynastic dictators to god-like proportions. A 100th birthday celebration would be a big celebration for them, worthy of something triumphant. Something game-changing even, like the successful launch of a satellite capable of holding a major weapon.

As I write this, online tracking shows KSM 3-2 is hovering above Amsterdam Island. This little French outpost is 10 kilometres long and 7 kilometres wide. It boasts a tiny population of around 25 scientists and researchers, all of whom are just visiting. It's right in the middle of the Indian Ocean – with Africa thousands of kilometres to its west and Australia equally as far to its east. From this vantage point there's little havoc this enemy satellite can wreak.

Traveling at the tremendous speed of 27,000 kilometres an hour, it takes a little over 90 minutes for it to complete its orbit around the earth. Every day it passes over major countries and cities whose people would likely feel a little uncomfortable to realize that the powerful toy of an unpredictable dictatorship was dangling ominously above them.

North Korea has certainly made it known it wants to cause major damage to its many declared enemies. Put the phrase "North Korea declares war..." into a search engine and watch all the suggested ways to finish that sentence. The government makes such statements so frequently that it's almost a national pastime. In January 2016, they claimed they'd successfully tested H-bombs that could destroy all of the United States at once. Then a month later Kim Jong-un declared nuclear war on China. The North Korean leaderships' ambitions know few bounds.

The majority of observers dismiss Jong-un's frequent threats as overblown rhetoric. They point out that the evidence shows North Korea's nuclear tests haven't been as successful as the country claims. While the basic facts of their nuclear aspirations and hostile intent are still alarming, the kilotons they've detonated so far aren't enough to cause the catastrophic damage they threaten. For the time being, it's all talk.

The common worry over the destructive effects of nuclear

12

weapons is that they will physically strike a target, causing a massive explosion that kills everyone in the radius. The atomic bomb that hit Hiroshima in 1945 had a blast radius of around 1.5 kilometres, immediately killing the 80,000 people in that direct area. A number of bombs stockpiled today are far stronger than the one detonated in Hiroshima. Yet the kiloton yields from North Korea's latest tests in September 2016 suggest they've barely caught up to the Hiroshima yield of 20 kilotons. So North Americans can sleep easy then, knowing that unstable dictatorships that declare war on us don't actually have the ability to seriously harm us?

Not at all. There is one major attack that North Korea could unleash on the continent. And they're not the only ones grasping for this nefarious tool. Several other countries and terrorist groups have just such an assault within their grasp. This attack isn't a bomb. It doesn't cause a blast. And it doesn't actually kill or immediately harm a single person. At least not in the very moment it strikes, that is. But it has the potential to cause more devastating consequences than the most powerful nuclear ground blast known to humanity.

What does it do? It turns off the power. This doesn't sound so scary at first, does it? After all, power outages happen from time to time. The kids turn on the flashlights, tell a couple of ghost stories, and then an hour later everything comes back on.

But what if it doesn't come back on? What if the power went off and stayed off all across Canada and the United States? For a couple weeks. Or a few months. Or even a year.

Then your food rots. The food at the grocery store rots. Everything at the regional food terminals that keeps our modern supply chain functioning goes bad. Water filtration plants shut down. Grandma's dialysis machine stops working. Cars and trucks don't start. A society built around electronics would simply cease to function. All with the flick of a switch.

This is what could happen if an enemy state or terrorist launched what's called an electromagnetic pulse attack – EMP – over North America. The effects would be beyond devastating. A U.S. government report even estimated that as much as 90% of the population could die within a year. Civilization as we know it would end.

On the same spring day that mysterious gunmen were shooting AK-47s at an electrical plant on the west coast, the KSM 3-2 satellite was flying over some of the most populous and militarily strategic parts of North America. It moved along a trajectory that flew over both Washington, DC, and the Toronto area. The so-called Bright Star sat at a perfect altitude from which to launch an EMP blast that would take out the electrical grid of the entire eastern seaboard and much of the Midwest or beyond. If it failed to take down the grid throughout the United States and Canada, more units of Metcalf-style gunmen could be strategically placed along the west coast to take down the remaining stations. Where could you reroute power to and from if the adjacent stations near and far were shot out or shut down? You couldn't. It would be lights out.

If the shots fired in the dark in Coyote Valley had been joined by another type of shot fired down from the dark of space, it would have been catastrophic. It's a big "if". The fact a still unsolved military-style assault on a power plant happened on the very day an enemy satellite flew at a perfect altitude and orbit to launch a crippling, coordinated attack could very well just be coincidence.

Let's just say it was. Let's say there was nothing bigger to that Metcalf attack. Even if that's the case, then what that day should have been was the biggest national security wake-up call in modern history. It wasn't though.

The Metcalf attack went largely ignored except by those who had the knowledge to fully understand the near miss of what could have been. Instead, most people have still never even heard of an EMP attack. Politicians hardly know about it, let alone know what to do to protect their countries from it. In the intelligence community, it's not even that widely discussed despite the fact it's one of the greatest threats faced today.

What are the chances this attack will happen? Hopefully very minor. We just don't know. What we do know with certainty is that Canada and the United States aren't protected against such a strike. One of the greatest threats imaginable is also our greatest vulnerability.

# THE CARRINGTON EVENT

The first major EMP attack in recorded history wasn't unleashed by North Korea, Iran or even a rogue terrorist group. In fact, it wasn't done by anyone on earth. And no, it wasn't aliens. The culprit for this serious assault that brought havoc all over the world was the Sun.

Most of us never give a second thought to our solar system's star as we go about our daily business. We look up to the sky each morning to decide whether or not we need to take an umbrella with us as we head out the door. We're certainly not accustomed to considering the massive sphere of hot plasma and gasses as a weapon of mass destruction. If anything, it's the opposite. In childhood, we're taught to view the Sun as a great giver of life. Toddlers sing rhymes about the Sun shining. School children place seeds in the soil to sprout by the classroom window so they can learn how plants take the earth's shining rays and transform them into energy via photosynthesis.

But the earth's greatest giver of life also has the ability to take that life away, or at the very least cause some major damage. Humans have feared the Sun's impressive powers as far back as recorded history goes. Many cultures, religions and mythologies make their Sun god their most awe-inspiring character. They know you don't mess with the Sun.

The story of Icarus in Greek mythology tells of a young man who discovers flight but ignores the warnings of his father to keep low. Instead he soars and flies too close to the Sun – burning his wings to pieces and falling to his death. The story is generally considered a parable about youthful hubris. But this tale proves that even thousands of years ago humans, with far less astronomical knowledge than we have today, knew the Sun had immense power over us and the earth. Powers that could lead to our demise.

We knew it back then. And we could, in some sense, feel it back then. But we hadn't yet experienced it. All that changed on September 2, 1859. On that day the Sun struck the earth with such power that it changed the way astronomers think about the over 4.5 billion year old yellow dwarf star around which we circle.

The day before was likely a fairly routine one for British astronomer Richard Carrington. For years he'd been studying the Sun from his private observatory some 30 kilometres south of London, England. Serious astronomy is a meticulous activity. For those who aren't passionate about it, tedious is probably a better word to explain the field. Astronomers can expect to spend many days spotting very little of interest to differentiate one day from the next. Carrington's notebooks are filled with such days.

On September 1, Carrington discovered something that made all those long hours of study worthwhile. He saw a cluster of sunspots – dark blotches on the atmosphere of the Sun. The exact nature of sunspots and their relationship to the Sun was still debated at that time. But Carrington at least knew enough to decide that the series of blotches he was observing was a big deal and made a detailed sketch of it. What happened next was even bigger and unprecedented in recorded astronomy.

A couple bright white lights appeared among the sunspots.

Carrington first thought there was something wrong with his equipment, that perhaps another light source was bleeding onto what he was seeing and altering it. He quickly checked his instruments and found that, no, this was real. Carrington didn't know what the lights were. He'd never seen anything like them. He watched with amazement until, only a few minutes later, the lights faded then, moments after that, completely disappeared.

Naturally excited with his findings, he went about making records, drawings and calculations. This was cutting edge astronomy - the kind of work that would ensure his place in the history books. He was already thinking about how he'd present his findings to his colleagues at the Royal Astronomical Society. He no doubt went to bed that night excited but completely clueless about what had really happened. It never occurred to him that what he had witnessed was the Sun launching a giant weapon of mass destruction in the earth's direction. It was about to strike the planet.

Eighteen hours after Carrington first observed his unique find, after he'd gone to bed and already got a good start on the next day, the sky changed. It turned different shades of red. At times it even shifted into blues and yellows. And it wasn't just in the south of England, close to Carrington's observatory, that this was happening. All over the world people remarked on the astonishing light show that they were experiencing throughout the day.

Diaries, letters and scientific notebooks recorded people's observations from as geographically diverse locations as Canada, Europe, Russia and South America. Several noted the sky was so bright they read printed text at a time of night that was usually pitch black. Unlike today, these observers weren't able to instantaneously share notes and corroborate stories across the globe. Yet they all recorded similar occurrences. More than one eyewitness report likened it to the sky being on fire. The world was experiencing what people close to the Arctic know as the northern lights, the aurora borealis, but on a much grander scale.

This was more than a pretty light show though. It was dangerous. Rugged sailors contended with bright red lights as they struggled through vicious storms. It almost seemed, some noted in surviving diary entries, that their ship was being sent on a journey into hell itself, with the sky ablaze and the waters roiling. Elsewhere, magnetic instruments went haywire, confusing travelers on land and sea. And whatever was happening interfered with telegraphs all around the globe. The communications devices sparked and caught fire. Operators rushed to turn off their machines and ran to find fire extinguishers or buckets of water. In some cases, operators were severely shocked by their own machines.

Canadian scientist D.H. Boteler has compiled many of the notes and diary entries people made in a 2006 paper for the journal Advances in Space Research. "Immediately I received a very severe electric shock, which stunned me for an instant," writes a Frederick W. Royce, a telegraph operator in Washington, DC. Over in Pittsburgh, telegraph manager E.W. Culgan witnessed "not only sparks (that do not appear in the normal condition of a working line) but at intervals regular streams of fire". Over in Norway, Professor Christoph Hansteen was watching telegraph lines and noted "pieces of paper were set on fire by the sparks of these discharges... it was necessary to connect the lines with the order in order to save the apparatus from destruction."

Many of these systems went down across the world for several days. This was a major problem because at the time telegraphs were the only way to communicate quickly. They were certainly the only way to get a message across a vast body of water, aside from delivering it via boat. Radio silence then was more than a minor inconvenience. Governments and businesses relied upon telegraphs for many of their daily functions. Suddenly they found themselves having to do without that information. The sunspots and lights that Carrington observed the day before shut down the 19th century's version of the information superhighway.

We now know that the Carrington Event was a geomagnetic superstorm that enveloped the earth's atmosphere in powerful energy. In and of itself, this is not necessarily a cause for alarm. The northern lights are a regular and harmless phenomenon also caused by magnetic energy. They're a small-scale version of what happened in 1859. But the earth's atmosphere, our shield, has limits to how much of a hit it can withstand. It can only absorb so much magnetic energy before it gives way under the strain and some of it leaks down to the earth, potentially causing harm.

Those flashes of light Carrington saw in his telescope on September 1 were in fact solar flares. These bright flashes are what occur when a burst of energy shoots out from the Sun's surface and hurtles into space. But they're not just made up of light. In a solar flare all sorts of particles and rays come along for the ride. The energy they release on average is equivalent to around 1 billion megatons of trinitrotoluene. Otherwise known as the explosive material TNT. To put it into perspective, the entire arsenal of nuclear weapons stockpiled across the earth contains only about 7000 megatons of TNT.

But what caused the Carrington Event wasn't a mere solar flare. It was a coronal mass ejection. This is a solar flare on steroids. It's what happens when incomprehensibly hot plasma from the Sun's atmosphere flies off into space. It took about 18 hours for the coronal mass ejection of 1859 to make it to earth. While they move slower than solar flares, this was still good time. After all, the Sun is 150 million kilometres from earth. This means it traveled at a speed of almost over 10 million kilometres an hour.

This might sound somewhat familiar to Canadians. That's because a similar but smaller solar event damaged the country only a couple decades ago. Luckily the impact was minimal. Back in March 1989 a whopper of a geomagnetic storm was brewing. A few days beforehand, astronomers observed a solar flare. Then they spotted the coronal mass ejection headed towards our planet. After a journey of a couple days, it hit earth and just like during Carrington's day, people all over the world saw unusual light displays in the sky.

"On 5 March 1989, complex sunspot region on the sun rotated into view from the earth," reads a study of the incident by scientists Sebastian Guillon, Patrick Toner, Louis Gibson and David Boteler in the September 2016 edition of IEEE power & energy magazine. "Over the next two weeks, this region was the source of numerous solar flares and plasma eruptions. Early and late in the sequence, eruptions shot out 'sideways' from the sun and missed the Earth. However, starting on 9 March, with the sunspot region now located closer to the center of the solar disk, a series of solar flares provided the first warnings that eruptions of plasma from the sun's corona were headed toward Earth." It reads just like accounts of the Carrington Event. This sequence of events "in turn induced electric fields in the power transmission lines that sent surges of electric current through the Hydro-Quebec power system".

The international political climate was tense at this time and imaginations ran wild. There was still another six months to go until Mikhail Gorbachev and George H. W. Bush declared the Cold War officially over. Many people feared the red lights in the sky were a precursor to a nuclear attack on North America. While the truth was less scary, it still made an impact.

This time no telegraph operators were shocked or burned by their machines. Instead, the surge of energy caused Quebec's power grid to shut down for nine hours. Families, businesses and government in Canada's second most populous province were left in the dark. Power was rerouted from other provinces and nearby U.S. states. A couple of pockets of the northeastern U.S. were also impacted. Even such a brief shut down was later calculated to have cost the

Canadian economy $6 billion. Power wasn't all that was affected though. Phone lines and other methods of communication took a hit, just like back in 1859.

The effects of the storm weren't isolated to that region of North America either. Australian forces operating in Africa as part of a UN peacekeeping operation even experienced the effects firsthand. Namibia was occupied by South Africa for decades but, after years of conflict, finally gained independence. A transition group formed by the UN monitored the first ever elections and kept the peace. The very same storm that hit Quebec rendered the Australian forces' high-frequency radio communications useless for a couple of weeks, putting a damper on their work. It's a minor incident but shows how geographically diverse the effects of such storms can be.

The disruptions around the world were otherwise brief and contained. The Quebec outage was the biggest problem. And it didn't even cause a cascading blackout, where a failure at one point along North America's eastern or western seaboards can trigger shutdowns at other locations along the coast. Or all of them. The North American electrical grid remained largely intact. Yet it still offered a sense of the awe-inspiring power a geomagnetic storm holds over modern society.

This is why solar astronomers and experts at research institutes such as NASA pay a lot of attention to the Sun's activities. They want to know as much as possible about how the Sun behaves and need as much lead time as they can get for any future coronal mass ejections that come our way. And there most likely are more to come.

Solar flares happen frequently, ranging from a few a day to one every week. Coronal mass ejections happen less frequently than that. Astronomers have figured out that the Sun has a schedule - a solar cycle - of about 11 years. It's during the solar maximum in this cycle, which lasts for a couple of years, that flares happen often and with the greatest ferocity. Scientists have compared the solar minimum - the opposite of the maximum - to something like being in hibernation. This is when we're least likely to get hit.

The last such maximum peaked in 2012, which means we dodged the bullet once again and can sleep easier until around 2023. But not too easy. One physicist has put the odds of a coronal mass ejection hitting the earth at 12% every cycle, a figure that NASA cites in their own predictions. Although that figure by no means guarantees a hit to earth in any given decade it's still too high for comfort given the potential consequences of a major impact. With a more than a one in ten chance, the odds are much, much better that a North American living today will see a solar flare strike than it is they'll win the lottery.

Besides, during that very solar maximum in 2012 one coronal mass ejection almost did smack the earth. It missed by only a week. It passed right through the earth's 150 million kilometre orbit. And it happened in July 2012, when the public and news organizations were more focused on the lead-up to the London Summer Olympics than they were in discussing the finicky but dangerous mannerisms of the dwarf star above that provided natural lighting for the games.

While this near-miss generated few news headlines right when it happened, it sure freaked out the science community. Many specialty articles, papers and studies went to print shortly after the event. Many of them pointed out how bad this could have been and how lucky we are to have avoided it. A seven day miss in an 11 year solar cycle calendar is too close for comfort. Especially given the significant power scientists estimated that one ejection contained.

The Quebec storm was calculated as having the power of Dst=-600 nT. This isn't something typically taught in high school science class. But it means the "disturbance - storm time" was minus 600 nanoTesla. The more negative the number, the crazier the storm. The Carrington Event is estimated at being anywhere from somewhat worse than the Quebec one to three times as bad. And the July 2012 one would have been twice as bad as the 1989 storm. Forget just the power grid going down for less than 12 hours in a province of home to 8 million people. Systems could have gone down in many provinces, states and countries around the world and for longer than nine days. Other electronics could have flared up like the old-fashioned telegraphs did. Electronics we rely on to keep our society functioning.

A year later, the UK financial institution Lloyd's of London teamed up with the American Atmospheric and Environment Research group to find out what the Carrington Event would cost the economy today. America alone would lose up to a whopping $2.6 trillion. That's more than 10% of the entire U.S. economy and more than the entirety of the Canadian one. It's clear how much the North America economy depends on having the energy grid up and running. Few other things work without it.

Put all of this together and it's easy to understand why experts are worried about coronal mass ejections colliding with the earth's atmosphere. They're basically naturally occurring EMPs. The good news is a sizeable number of satellites, telescopes and astronomers are always keeping an eye on the Sun. If the Sun awakens from its slumber and another flare up happens, they'll know in advance. We'll get a day or two of warning to prepare. We should be able to calculate its force and speed and time of arrival right away.

This means countries and industries can prepare their infrastructure and enact measures to protect people from any harmful impacts. The emergency measures some observers recommend include powering down our grids and getting our aircrafts to fly at lower altitudes. "The operating strategies can consist of postponing routine maintenance or putting equipment back into service so that the system is in its most robust condition before the storm hits," a paper on the effects of solar storms co-produced by Natural Resources Canada and the Finnish Meteorological Institute in Finland advises. The more severe storms will require even more preventative steps. This won't stop every negative consequence from hitting home, especially since some of the effects are still unclear, but it'll mitigate a whole lot of the hurt.

What if we didn't have any warning though? What if there was just no time to prepare? What if an EMP blast went off in the atmosphere and nobody knew about it until it was too late? You'd only figure out what happened after the storm hit. By that time, the damage is done.

This is exactly what would happen if an enemy decided to launch an EMP attack. And it wouldn't be coming from outside of the atmosphere, where its strength could be diluted. It would come from within. And it wouldn't be randomly positioned either. It would be targeted, to make sure it rained down on the intended victims. It's like harnessing the powers of a coronal mass ejection and using it as a weapon, but with far worse consequences.

Nobody was seriously considering the possibilities of this in the aftermath of the Carrington Event. For many years the event was mostly fodder for solar astronomy and the scientific community. The thought of any villainous application didn't even cross the minds of anybody in North America for decades to come. All that changed though when American government scientists and military experts were toiling away on a deserted island far off the coast of Hawaii in the 1960s. There they happened upon an unexpected and alarming phenomenon, one that opened their eyes to the possibility of man-made EMP attacks.

# STARFISH PRIME AND HAWAII

Ferdinand Avenue sits to the northeast of downtown Honolulu. It's located in a pleasant upper middle class area, part of the Manoa neighbourhood, where house prices comfortably cross the million dollar mark. Its palm trees and spacious lots depict the sort of relaxed scene you'd expect to find in Hawaii's capital. Yet it was also the site of a curious incident as equally important to this story as the case of the Metcalf transformer attack. On July 8, 1962, shortly after 11 p.m. the street lamps went off on Ferdinand Ave. and a couple of connecting streets. Thirty street lamps, to be exact.

Normally, such a localized blackout like this would be of little significance, if any. Maybe it was caused by a faulty part that was nearing the end of its life cycle? Or from damage caused by the island's troublesome boiga irregularis – a brown tree snake – slithering up the wires? Not this time though. This case was different.

And anyone who can authoritatively explain the Hawaiian streetlight incident can in turn explain what EMP is and why it poses such a threat to our way of life. But first it's important to know why anyone aside from a Honolulu city maintenance worker would even know about the Ferdinand Ave. lights going out.

In the early 1960s the Cold War was well under way. Both the United States and the Soviet Union wanted to keep learning as much about nuclear weapons as possible. And they were both regularly conducting experiments – blowing up bombs to learn about things like the size of their detonations and the amount of nuclear fallout. Even though the U.S. had dropped two nuclear bombs on Japan in August 1945 as part of World War II, they still knew very little about the potential uses and effects of nuclear detonations.

The secretive Manhattan Project was created by the U.S. government in 1942 to take the theory of nuclear fission discovered by the Germans in 1938 and devise a weapon out of it. In many respects it was a hasty affair. They conducted the first ever nuclear detonation in history – the Trinity test in New Mexico – only several weeks before they dropped the bombs in Japan.

While it's clearly good for humanity that there have only ever been two nuclear bombs used in warfare, it makes for scant primary source data for nuclear scientists to study. Despite the billions of dollars spent on research and the tens of thousands of employees working for the Manhattan Project, they were still making it up as they went along. And even after the project shuttered its doors in 1947, nuclear scientists still resorted to using guesswork to help them figure out this emerging field.

Luckily more data was to come their way. As the nuclear arms race surged so did the number of nuclear tests. 1962 was a busy year for both the USSR and America. The Soviets conducted 79 tests and the Americans fired off 96. For both countries, these were by far their greatest number of tests during one calendar year. It was also the year of the Cuban missile crisis – the year when Americans feared for their future and school children were taught to hide under their desks in duck-and-cover drills. So given everything that was happening, it's understandable that the news and rumours of a seemingly minor but exceedingly important new discovery about the effects of nuclear detonations made during one of the almost 100 tests that the United States conducted that year never made its way to politicians, the press, the public and even most people in the military and scientific communities.

Operation Dominic was put together quickly against the background of a changing political scene. In 1959 and 1960 neither the Americans nor Soviets conducted a single test, abiding by a moratorium. But in 1961 Soviet premier Nikita Khrushchev decided to drop the ban and the USSR started testing again. The Americans responded in kind and President John F. Kennedy authorized the Dominic series of tests. The bulk of this series involved dropping 31 bombs on or around Christmas Island, which sits in the middle of the Pacific Ocean. Every few days, mostly during May and June, there was a new test.

The bombs were dropped from B-52 aircrafts either in free-fall or with parachutes attached. The main purpose of the tests was to determine the success and effects of the detonations. As sad as it sounds, if you're going to drop a nuke on another country you want to make sure you at least do it right. This basic on-the-ground explosion, and the ensuing crater damage it causes, is generally what comes to mind when people think about nuclear bombs and their impact. However Operation Dominic had a subsection of tests with a different set of goals. Operation Fishbowl, as the 10 additional tests were called, didn't drop bombs from a plane onto the ground. They shot them up into the air. Way up.

High-altitude nuclear detonations, as they're called, are when a warhead is attached to a missile and then fired many kilometres up into the sky. It's risky business. It's not uncommon for such nuclear launches to run into trouble. They're tricky to get right. They can fly off course or the rocket can fail. These are tightly controlled tests so if they don't go according to plan, officers on the ground will order the warhead to be safely destroyed before detonation. It means they have to start all over again, but it's better to be safe than sorry. And the first three intended Fishbowl detonations – Urraca, Bluegill and Starfish – all ran into trouble and never made it to the final step. The next up was Starfish Prime, which was a second attempt at the original Starfish launch.

For the launches, U.S. naval officers and scientists gathered on and around Johnston Island, the main portion of the Johnston Atoll in the North Pacific Ocean. An atoll is a group of islands formed by coral. The island is technically U.S. territory, but despite its tropical climate nobody has ever lived on it. Its only residents over the years were military personnel on duty. It's too far out in the middle of nowhere for anyone to want to live there, which is the very reason it's the perfect place to conduct nuclear weapons testing.

The failed Starfish launch was on June 20, 1962 and by July 8 the Fishbowl team was ready to have another go at it. But what's the point of conducting tests way up in the sky like this? It's not like they can be used as weapons in war. At least that's not what any scientists thought at the time. They did however have an inkling that something different happened during high-altitude blasts that didn't occur during ground detonations. For as long as they'd been investigating and testing nuclear weapons since the 1940s, physicists were certain high-energy electromagnetic waves were released by nuclear explosions and interacted with the atmosphere. They knew it happened mostly up in the air. And they knew this phenomenon interacted with electronic circuits and did things like disrupt radio waves. The degree of the disruptions was unclear though. They just initially thought the effects were minimal and warranted little cause for concern.

A planning document for Operation Fishbowl dated November 1961 produced by the Air Force Special Weapons Center shows that as the years went by they began to have thoughts to the contrary. Concerns about a damaging electromagnetic pulse were growing. Maybe the effects were worse than they thought. The phenomenon warranted further investigation and the Starfish launch was conceived.

"The purpose of this test is to determine the intensity and duration of the ionized layers produced in the upper atmosphere by the deposition of bomb fission debris and radiations, and to determine the efficiency of injection of relativistic electrons into stable orbits in the earth's magnetic field," the report explains. "These effects produce intense and persistent upper atmosphere ionized layers, both in burst region and at the magnetic conjugate point, which cause radio and radar blackout over large areas. In order to achieve sufficient understanding of phenomenology of a nuclear burst at this altitude and to permit accurate prediction of system blackout effects, the following test observations must be made".

It's a lot of technical jargon. But what it proves is that the U.S. knew that one of the effects of nuclear detonations was the release of unseen electromagnetic energy that travels like waves, like a pulse. They knew this because they'd observed versions of it in tests during 1958 but also because physicists had worked it out on paper. Too many questions still lingered. Ones that needed to be answered, and urgently, given the pace of the arms race. Does the pulse shut off their communications systems? Does it scramble the radars of their fighter jets? And how severely? And for how long? And under what circumstances? They needed to launch more high-altitude tests if they wanted to get to the bottom of this emerging concern.

The beginning of that report, which has since been made public, states: "This document is classified SECRET – RESTRICTED DATA because it contains information on nuclear effects." They add: "The information contained in this document will not be disclosed to foreign nationals or their representatives." No big surprise there. And anyone in the U.S. government who found this report on their desks would know that was a given. Of course they didn't want other countries to know about it. Especially their enemies. They were onto something, but hadn't yet figured it out. It's a vulnerable position to be in. They knew enough to know there was trouble on the horizon, but few details about that trouble. If the enemy knew it before them, they'd gain a tactical advantage.

These were the thoughts that occupied scientists and military personnel as they watched the Starfish Prime launch shoot up into the sky on that summer night. The actual projectile was a 1.4 megaton nuclear warhead strapped onto a Thor missile. This is an intermediate-range ballistic missile that can travel over 3000 kilometres. And unlike the Fishbowl launches before it, this missile was a success. The rocket shot up over 1000 kilometres before it hit its apogee. That sounds like it's pretty high up. Into outer space perhaps. Though it's technically not. It's still well within the exosphere, which is the outermost part of the earth's atmosphere, where gravity still holds sway and there is a corona visible from outer space. Although it did fly a bit higher than where the International Space Station currently orbits, at slightly above 400 kilometres.

Still, all of this is to say that when an EMP-enabled warhead goes way up to view its potential victims, it's got a wide view of possible targets to choose from. The Thor missile more than did its job. The 1000 kilometres far exceeded the desired detonation height for the warhead. It wasn't until the missile was on its downwards trajectory and hit 400 kilometres altitude that the warhead was detonated. That was the moment when researchers finally got what they wanted. The burst was a success and an EMP was generated. The scientists got their much coveted data from this launch, which is information that is still used in calculations today. The trouble is there wasn't as much data gleaned from Starfish Prime as there could have been. This is because the EMP was so much larger than expected that eagle-eyed observers and their equipment were oriented to look for different readings and results. They committed to be even more prepared for the next test, where they'd build on their already increasing knowledge base.

This is a good place to pause and seek a good explanation of what EMP actually is and what it does, in a technical sense. A fair amount of ink has been spilled on it over the decades - even if it still hasn't captured the attention of the public, media and politicians. Scientists certainly calculated many numbers and observed a lot of effects. How exactly do they describe this thing? It's a challenge to find a decent description written in plain English in the research findings, reports and papers if only because scientists don't always write about the phenomena they're dealing with by using words, but by equations. It's kind of an insider lingo that only specialists can understand and is of little use to the average person.

The most succinct explanation comes from a 2004 report by the EMP Commission, which was established by U.S. Congress: "Gamma rays from a high-altitude nuclear detonation interact with the atmosphere to produce a radio-frequency wave of unique, spatially varying intensity that covers everything within line-of-sight of the explosion's centre point."

A slightly more technical but still readable explanation comes from Conrad Longmire, whose calculations from the 1980s confirming the science of high-altitude EMPs are still considered the standard today. "Nuclear bombs emit a small fraction, of the order of 0.003, of their energy in gamma rays," Longmire wrote in a 1986 study. "The principal interaction of gamma rays with air atoms, or other matter, is Compton scattering. In this process, the gamma collides with an electron in the air atom and knocks it out of the atom. In so doing, the gamma transfers part of its energy (on average about half) to the electron, and is scattered into a new direction. The Compton recoil electron goes generally near the forward direction of the original gamma, never in backward directions. Thus a directed flux of gammas produces a directed electric current of Compton recoil electrons. This current produces the EMP."

What does this all mean then? Perhaps most importantly, it makes it clear that for all the big explosions happening 400 kilometres up in the sky and for all the intimidating looking naval aircrafts clustered around the Johnston Atoll and for all the bluster of the arms race, the part of this everyone should have been concerned about wasn't any of those Cold War images but the fact that 30 street lamps went off in a sleepy suburban community over a thousand kilometres away.

That was the biggest alarm bell that came out of Starfish Prime. It confirms that this fancy "directed electric current of Compton recoil electrons" Longmire describes is something nobody would want passing through their electronics. The Hawaiian streetlight incident was the first piece of tangible evidence suggesting that EMP could cause large scale damage to not just military radars but civilian infrastructure. And it made it clear that EMP wasn't just a passive effect resulting from geomagnetic storms, but something that can actually be created by human beings and used at the time and place of their choice. It was a man-made weapon.

If the Fishbowl detonations had been conducted closer to civilian communities there might have been much more evidence of civilian damage. But the government wanted to conduct the series of tests as far away from people's homes as possible. It's hard to find a U.S. territory further away from both mainland America and other nations as Johnston Island. Its closest inhabited neighbour is Hawaii – at 750 nautical miles away, or 1400 kilometres. If they could have found a suitable test site even further away, they surely would have taken it. This wasn't so much done for the sake of secrecy, although that certainly played a part. Rather, it was a high priority on the part of the government to shield civilians from any negative effects of the tests. And they were largely successful. The atmospheric blast didn't cause damage to anyone's homes. The released radiation didn't make any civilians sick. And the flash of light that comes with the blast, which has blinded people during past nuclear tests, didn't harm any Hawaiians' vision.

This lack of collateral damage is largely due to how researchers understood the potential hazards of typical nuclear detonations and made decisions aimed at preventing or minimizing them. But at this time they didn't know all that much about EMP. They didn't know just how far its reach extended. It's why the following line in the July 9, 1962 edition of the Honolulu Advertiser was a surprise to everyone: "The street lights on Ferdinand Street in Manoa and Kawainui Street in Kailua went out at the instant the bomb went off, according to several persons who called police last night".

Other news reports indicate that as many as 300 street lamps went out across Hawaii right at the time of the detonation. It's an alarming fact that a nuclear explosion 1400 kilometres to the west and 400 kilometres up in the sky could blow out street lamps. If that was the case, what other unexpected damaged could it cause? But scientists aren't going to read newspaper reports and take them as proof of cause and effect. How can anyone be sure these outages are connected to the nuclear detonation? They could just be coincidences.

This is why that strip of lights on Ferdinand is so important. The outages are largely hearsay but they're no doubt true. There's no reason anyone would lie about them. And there's a good chance they were caused by EMP. However when it comes to the Ferdinand black-out, this is the street where there's extensive recorded data for posterity to analyze as proof. This is because a physicist from Stanford University, John Mattox, personally visited the site to investigate. And not just to take a look at the lines. He interviewed the line foreman who repaired the lights that night. He spoke with the sales engineer who provided the parts that failed. He then inspected all the parts on the lamps to figure out various nitty-gritty details like what voltage ratings they had and what voltages would cause them to short circuit.

Researchers have used Mattox's data and information to

conclude the black-out wasn't caused by equipment failure. And it wasn't caused by a random lightning strike or power surge in the area. In 1989 physicist Charles Vittitoe, working for a U.S. Department of Energy research lab, took all of the technical information about the Ferdinand lamps and crunched them alongside all of the data about the electromagnetic energy sent hurtling down from the Starfish Prime blast and concluded "the estimated voltage is consistent with the observed effect being caused by EMP." Starfish Prime fried those lights.

Vittitoe also found a couple of key explanations as to why these lamps in particular went off. One was that some of the parts were nearing the upper limit for the voltage they could handle. It wasn't going to take much to disable them. Another was that the actual circuits within them were facing the direction of the EMP and lined up so the pulse hit them at just the right angle. This is why the definition from that 2004 report to Congress notes that EMP can knock out "everything within line-of-sight of the explosion's centre point." Starfish Prime blew up mostly out of sight from civilization. But not entirely. The pulse doesn't just shoot directly down and impact whatever is below. It moves out diagonally. The blast's radius is much, much wider than the radius of the actual detonation. It's a key point to understanding its potential as a weapon of mass destruction.

What if the blast had gone off directly above Hawaii? Or above New York City? Would many of the street lamps in those cities have gone out? Would other electronic devices? It was another question researchers wanted to answer. But the area the pulse covers isn't the only variable that, if changed, causes trouble. Vittitoe concludes that "a faster rise in the EMP, a higher peak EMP value, or a further delay in arrival of reflected signals would have caused increased voltage and produced more failures." In other words, a stronger EMP combined with less sturdy electrical parts will wreak much greater havoc.

The expected response from the naturally curious scientific

community was of course to want to learn just how much more trouble an EMP causes. Let's do more testing out in the middle of nowhere, they'd say. But maybe this time we should put some other street lamps or electronics on a barge or other little islands and see how everything interacts with the EMP. Let's find out more about this as soon as possible. Then let's work to protect ourselves from any negative effects. And let's do all this before our enemies know more than we do and wind up with a tactical advantage over us. Kind of an obvious plan, no? Sure. But there was a little problem.

It all came to a halt. The U.S. signed a treaty that ended high-altitude nuclear testing. Following Starfish Prime there were two other detonations in the Fishbowl series - Bluegill Triple Prime and Kingfish - that yielded strong EMP waveform data. Some of this information remains hidden to this day. However experts have let slip that this data at least confirms their understanding of EMP as it relates to phenomenon like the Hawaiian streetlight incident.

But after Kingfish, that was it. On August 5, 1963, the U.S., USSR and United Kingdom signed the Partial Test Ban Treaty. It prohibited all types of nuclear tests aside from those done underground. The signatories' goal was to slow down the arms race while simultaneously reducing the amount of radioactive dust in the atmosphere. No more bombs going off in the desert, like the Trinity test. No more detonations in the air. Now all they could do was blow them up in underground shafts. And the U.S. certainly did just that. Operation Storax got underway before the Fishbowl series was even complete. They performed dozens of such detonations - sometimes more than one a day. No doubt they learned a lot from them. The only problem was there was no EMP. Gamma rays can't react with the atmosphere when there's no atmosphere with which to react. The evidence gathering had come to an end.

It was an unfortunately ironic situation. Just at the time physicists were really getting to the bottom of a serious and dangerous effect of nuclear detonations, their tests were banned. No more observations of EMP. It was a good move. In some respects. The test ban was a positive step forward for world peace and nuclear non-proliferation. But it didn't help science. In the decades since the signing of the test ban, research in nuclear physics has plummeted. There's less public interest. This means there's less financing available, both public and private. This in turn means there's just less of a motivation for intelligent and ambitious researchers to enter this specialty. The result is a knowledge deficit in the continued understanding of nuclear physics. Do we even have answers to those questions that Vittitoe and others like him put forward? The warning he gives is a sombre one. It tells us that if our technology changes, the reaction to EMP also changes. And one of the things that changed the most between 1963 and today is technology. This leaves one big question: Are we now more or less vulnerable in the face of an EMP attack?

# THE PARADOX OF THE ELECTRONIC CIVILIZATION

For a few months in 2000, Sunshine Coast lost its lustre. The east-coast Australian city is almost always paradise. Its summer temperatures regularly hover in the high 20 degrees Celsius. And even in winter they rarely dip below 20 degrees. It's no wonder its clean waters draw surfers and sun-worshippers from across the world. Of all the locations then for an act of sabotage known as the Maroochy water breach to happen, Sunshine Coast was one of the worst. The incident quite literally stunk up the town.

"Marine life died, the creek water turned black and the stench was unbearable for residents," an Australian Environmental Protection Agency told media at the time. The Maroochy Water Service - the municipal water company that was attacked - had safeguards in its computer systems to protect such disturbances from happening. It was a properly managed and regulated facility, like most public utilities in developed nations. But for several weeks beginning that March, random technical failures cropped up without rhyme or reason. Pumps ran for longer than their programming dictated. Or sometimes they just shut off randomly. Then unexpected and increased radio traffic at the station added to system failures by confusing signals.

Alarms randomly went off. And because of these seemingly minor malfunctions over one million litres of untreated sewage was dumped into local waterways, dimming the city's shine.

Engineers conducted test after test to find the cause of the problems. Management rebooted and reinstalled the software that controlled the systems. They tried every internal intervention that came to mind. But the hiccups continued. Eventually they discovered the culprit: sabotage. It became clear a hacker was at work disabling the water systems that Sunshine Coast's 50,000 residents relied upon. At one point, a company employee working online even came across the phantom presence of the enemy in cyberspace, as they both toggled with the same controls at the same time. Finally the police got their guy. And it wasn't a terrorist or some foreign saboteur.

In 2001, 49-year-old Vitek Boden was given a two-year jail sentence for hacking into the Maroochy Shire waste management system. His motive wasn't a desire to poison the water supply or wreak ecological damage. He'd orchestrated the whole messy deed because his application for a full-time job with the town council was rejected. That's it. He was nothing more than a disgruntled aspiring employee. Yet because he'd recently worked within that very water plant for a contractor that had serviced the system, he knew how to access its controls from his laptop. It was a far more elaborate plan than slashing the boss's car tires in frustration.

This isn't just a local story about an isolated hacking incident. Neither is it a tale about the importance of human resources etiquette when rejecting a candidate. At least it isn't for the cyber-security and infrastructure experts who keenly watched and studied the Maroochy water breach as a sign of a larger problem. Seen through their lens, Boden's story is a key example of a problem with our electronic civilization. That's the term used to explain how more and more facets of our daily lives are governed and controlled by interconnected electronics.

While some of our dependence on electronic devices - automobiles and refrigerators, for example - is obvious, the other devices we rely on are less so, but no less important. SCADAs fit the latter category. These little known devices are vital to contemporary living. SCADAs are what Vitek Boden - a down-in-the-dumps unemployed technician - played havoc with. Sitting outside the waste management plant with his laptop, he batted them about like a cat with a mouse.

The acronym stands for Supervisory Control And Data Acquisition. It's not the most eloquent term. It's also not particularly useful, as it still doesn't really explain what it does. "Supervisory control and data acquisition (SCADA) systems are widely used to monitor and control operations in electrical power distribution facilities, oil and gas pipelines, water distribution systems and sewage treatment plants," write Jill Slay, an engineering professing at the University of South Australia, and Michael Miller, from the defence firm BAE Systems, in a 2008 conference paper titled Lessons Learned from the Maroochy Water Breach. Slay and Miller saw what the media and most other observers missed about the Maroochy water system – that it highlighted a major corporate and national security weakness.

SCADAs are modern equivalents of the Roman aqueducts – the ancient tunnels engineered to bring water to towns and cities. They manage, as Slay and Miller point out, the safe and steady flow of pretty much every major public and corporate utility. For the labyrinthine sewer systems to work in major cities, the water can't just flow willy-nilly. What happens when one part of the city dumps significantly more sewage into the system at one time than another? The inconsistencies create chaos if they're not managed. There needs to be some form of regulation. Likewise with electricity and oil and gas.

A 1949 video stored on the Getty Images archives shows

engineers at a power plant in Perthshire, Scotland, walking about the station and opening and closing giant valves. They're manually regulating the flow of water and electricity. By monitoring the various levels and comparing them to their charts that list the optimal volumes, they know when it's time to open and close the different spouts and currents.

Decades ago this was a manual labour job. Now, in the age of the electronic civilization, things are done differently. Blue-collar jobs at factory assembly lines aren't the only type of work that is now automated. The safe regulation of public utilities has gone that way as well. SCADAs are programmed to read various outputs, cross check them with the set levels and then send out messages to all the different spouts and spigots and electrical controls instructing them to make adjustments. Nowadays what workers on the floor do is simply monitor the monitors – checking in occasionally on the SCADAs and making sure they're doing their job. If not, engineers reprogram them or replace their faulty parts.

Initially, the automation of major utilities – the arteries of modern civilization – wasn't considered much of a security threat. As long as SCADAs are maintained and monitored, they work just as effectively or more so than manual operators. After all, there's no chance for human error. If somebody wanted to mess with the system, they'd have to tamper with the controls on the SCADAs or in a control room manually.

But that's not what Vitek Boden did. He interfered with the system from outside of the plant, using only his laptop. He was able to do this because the electronic civilization is also now the interconnected civilization. The Maroochy water service was online, connected to the internet. In addition to the electronic civilization, we live in a time where more and more devices are a part of "the internet of things". The term explains how it's not just computers and phones that are connected to the internet. Now you can connect the baby monitor to the web to check in on what the babysitter is getting up to. You can connect your thermostat to WiFi, allowing you to adjust the temperature in your home from, say, your workplace. The problem is if you can tinker with your appliances away from home, other people can too. Security is far from air tight. Someone who knows what to look for and how to get into the system can potentially hack into it and control your baby monitor. Or something larger. Like a regional power grid. Or water flow. And that person's nefarious ambitions might not be limited to getting revenge on a local water system for a lost job. They might be thinking big picture and aiming for nation-wide mayhem.

"The incident was serious, but it was caused by a lone hacker who attacked just one system in a single infrastructure," Slay and Miller's study concludes. "One can only imagine – at a time when terror threat levels are high – how much devastation could result from large-scale coordinated attacks on SCADA systems throughout the interconnected critical infrastructure sectors."

This is a scary prospect. But when it comes to the liabilities of hacking, the experts are all over it. Open any technology magazine and you'll find an article about cyber-security. Governments around the world have convened committees and created departmental divisions to deal with the emerging threat. They've appropriated funds and passed legislation. Private contractors offer their expert services to large corporations. Tech savvy consumers download protective software for home use. The world is aware of the many risks stemming from hackers and other such cyber attacks. There's usually a plan in place, even if, as in the case of the Maroochy water service, it's not a very effective one. You can't say the same for EMP attacks though.

There is no plan. No awareness. Nothing even coming close to the degree of attention the cyber-security world enjoys. And if SCADAs are already this vulnerable to minor league threats from hackers, when it comes to much bigger threats like an EMP, they don't stand a chance.

For a better understanding of how EMP attacks can take down SCADAs, let's return to what happens when a pulse encompasses a geographic region. While it impacts all electronics in its sight, there are some things it won't harm. It'll travel through glass and wood structures unchanged, sending its damaging high voltage into whatever electronics it encounters and frying them. But there are certain metals it doesn't pass through.

SCADAs, as a physical object, are usually a collection of computer parts – kind of like the motherboards in desktop computers – that are typically housed in a metal box. Every day we pass these sorts of little boxes – often grey or green coloured – that have something to do with the phone company or the electric utility and we never think twice about them. SCADAs are just like these - and some of these we pass likely are SCADAs. Depending on factors like whether they're stored outdoors or in a basement and what sort of covers protect them, an EMP may or may not directly impact the SCADA box.

This is why experts recommend encasing each SCADA in metal. There are ways to significantly minimize the threat from the pulse, which we'll discuss in detail in a future chapter. Now this simple cautionary fact could serve to eliminate a great deal of worry about the negative effects of an EMP attack. Just cover important things in metal and we're all good, right? Wrong.

EMPs are energy. They're electricity. And electricity travels, whenever it has the opportunity. It doesn't like to stop, either. It keeps on flowing. Once a pulse gets into connected circuits it moves through live wires and enters businesses, homes, wall sockets, lightbulbs, computers, SCADAs... you name it. But a grid that is used to a controlled and regulated amount of voltage flowing through it suddenly gets overwhelmed by the high voltage from the EMP. The EMP current runs amok, overwhelming whatever's connected to the grid. This is how those telegraph machines caught fire and harmed their operators during the Carrington event.

Think of it along the lines of when a group of older kids show up uninvited at a house party. Things get out of control quickly. The place gets trashed.

The damaged SCADAs at the Maroochy water service illustrate the paradox of our electronic civilization. We rely entirely on electricity and electronics. Society is increasingly automated and interconnected. This is supposed to make life better. And in most aspects it clearly does. But this doesn't come without a massive liability.

Not that long ago, if you wanted to mess with the water system, you'd have to break into the control room and physically fight the blue collar workers tasked with looking after the valves and pumps. And good luck doing it hundreds of times in a row all in the same day with the goal of shutting down an entire city. Now all you need is a guy with a laptop who got up on the wrong side of the bed that morning. Thankfully most people can take rejection better than Vitek Boden. But it's not only our increasing reliance on technological automation that creates this paradox. Innovations in computer science are also making the electronic civilization more fragile.

"I'm going to put music in your pocket," Steve Jobs says to his daughter, Lisa, at the end of the 2015 movie named after the head of Apple. It's a visually striking scene – with father and daughter standing opposite each other on a rooftop, an urban vista of tall buildings surrounding them. The scene is meant to represent both the next step in their fractured relationship and the next big leap in technological innovation. Neither of them is carrying anything. Jobs, of course, wears his signature glasses. The only obvious prop is the clunky walkman strapped to the side of Lisa's pants. "A hundred songs," Jobs continues. "A thousand songs. Five hundred songs. Somewhere between five hundred and a thousand songs. Right in your pocket. Because I can't stand looking at that inexplicable walkman anymore. You're carrying around a brick playing cassette tape."

The scene takes place in 1998. Three years later, Jobs released the first generation iPod, which did fit right in a pocket and held enough music to fill a bunch of boxes in a garage. If the music was on LPs or cassettes and not digitized, that is. The whole iPod only weighs about 6.5 ounces. And a single clunky audio tape, when in its case, weighs just about half of that. Yet the iPod did more than carry twice the number of songs as a cassette tape. It carried around a hundred more. In the years since, that number has multiplied several times over and now an Apple device can carry enough tunes to fill an entire garage and then some. And the devices do much more than just play music, as anyone who's ever listened to music while doing their banking right after video-messaging with friends can attest.

The final scene in "Steve Jobs" is an optimistic one. As is almost every story about the development of such tech devices. They make our lives better, the narrative usually goes. And there are seemingly endless examples to back up this claim. It's true on the small scale, like how an app can remind a senior citizen when to take her pills and how many to take. It's true on the large scale, too, like how in 2011 journalists and activists in Egypt's Tahrir Square used social media to get instant alerts that helped them weave through the crowds and avoid government thugs trying to clamp down on the emerging revolution.

But this story doesn't come without its liabilities. It's not entirely feel-good. The fact that in the 21st century people do so much, so often, on such small devices presents its challenges. Some of which are more obvious than others. How long do new parents take pictures and videos of their newborns until it occurs to them to back them up on external hard drives or in cloud computing? If it takes them too long and the phone breaks or is stolen, say farewell to months or years worth of family memorabilia. If your phone holds all of your passwords and a hacker-turned-pickpocket comes a-calling, consider the doors to the virtual safe wide open.

It's also the increasingly small size of the devices themselves that cause significant problems. The compactness of Jobs' first iPod heralded a new age of convenience. There is no doubt about that. But the smaller the devices, the easier they are to damage. Way easier. If you spill a whole cup of coffee near a desktop computer, which fewer people and businesses now use, it's likely just the keyboard that'll need drying out. But that much liquid is enough to completely soak a handheld device – both inside and out. Good luck trying to get the manufacturers' warranty to pay for your clumsiness. Likewise if you elbow a desktop computer tower, it may barely budge. But if you drop a handheld device from even waist level, the damage – like a cracked screen – could be irreparable.

These are mere inconveniences though compared to contemporary technology's more troubling weaknesses. Today's electronic devices are much more fragile than they were decades ago. It's a fragility that makes our societies more vulnerable to attack. Particularly EMP.

While researchers first clued into the possibility of EMP attacks during World War II and the subsequent arms race, it likely didn't cross their minds that those electromagnetic waves could also take down computers. Why would it? The few computers that existed were some of the sturdiest pieces of electronics ever invented. The war was a period of innovation not just for nuclear science but computing as well. Necessity is the mother of all invention, as the saying goes, and in the 1940s the United States needed to quickly figure out how to more accurately fire missiles at enemy targets. These sorts of calculations were called artillery firing tables and could be worked out mathematically on paper by hand. It was a time consuming affair though and time was a luxury the Allies didn't have in the theatre of war. In 1942 the U.S. had a hundred people working six day weeks over two shifts at the University of Pennsylvania just to create a new set of firing tables. But even this toil was proving insufficient and they were worried they wouldn't produce the new tables before the war was over. It was with this problem in mind that researchers conceived of and created one of the world's first computers – the Electronic Numerical Integrator and Computer.

The ENIAC, as it's called, isn't just a marvel of programming, but of engineering and construction as well. It was a massive device – many metres long and reaching up close to the ceiling. The frame was made of steel. The builders ordered 26,000 feet of wire during construction. There were half a million metal joints in it that needed to be soldered together by a hot iron. And the number of vacuum tubes – the integral piece of hardware that makes the computations possible – is estimated at a whopping 18,000. This was a big, sturdy beast of a machine. There was no way you could knock this thing over. You'd have a hernia if you even tried to move it an inch. Spill your morning coffee over one of the thousands of vacuum tubes? Just replace it... if it even gets damaged.

It's not just when it comes to physical destruction that the ENIAC enjoys an advantage over today's pocket-sized devices though. One of the innovations that allowed the smart people at companies like Apple and IBM to build smaller devices is that they need fewer and smaller parts to do increasingly complex tasks. Resistors, capacitors, oscillators and all those tiny metal and plastic parts have also considerably advanced. They can pull off more complex and faster calculations than ever before. The computing speeds increased over the years at a rapid, often multiplying, pace. While all the ENIAC calculated was firing tables, the latest generation of handheld devices does so much more than the creators of their massive ancestor ever imagined.

There's a catch though. A big one. There's only so much electrical current that tiny electronic parts can handle before they shut down or become damaged. It would have taken a great deal of electricity to knock the ENIAC behemoth out of action. But it takes much less to fry an iPhone. Or any of the other many pieces of electronics we use in our daily lives. And not just the ones we use for mundane tasks. These sorts of improvements also changed the devices used by military personnel and police, medical experts, senior politicians and multinational business executives - the type of people whose work shouldn't be easily crippled. But now can be.

This compounds the paradox of our electronic civilization. It's not just that we rely on devices ranging from SCADAs to iPhones. It's that their robustness hasn't kept pace with their complexity. Developments we consider improvements have also made them more fragile and vulnerable to attack. Collective innovations have made life much easier and more comfortable for many human beings around the world. Clean drinking water. Sewage systems. Air conditioning. Indoor lighting. Refrigerators. Microwaves. Automobiles. Washing machines. Life without these inventions would be unthinkable for people accustomed to modern 21st century living. Yet what do they all have in common? They all rely on electronic components to function. If the electric grid goes down, if a powerful EMP shoots through wires and power cords, if a significant percentage of our electronics are dismantled, our way of life is shut down.

It wasn't always this way. In the past, it was near impossible to topple a broad civilization with a single effort. When people used to get their drinking water from a fresh body of water and cook their food over fires they made out of wood they gathered, there was no connection between being able to drink and eat. There was no single problem that could arise that would make both of these life sustaining activities inaccessible at the same time. And to stop people from accessing water back then, you'd need to dam up the waterways or pull off some equally time consuming and logistically challenging effort. To stop them from cooking their food, you'd need to cut down every tree or control the rainfall so it put out their fires. Yet nowadays if the power goes out, not only do the water pumps and sanitation systems fail, so does the refrigerator, stove and oven. The electronic civilization has chained us to the electric grid and everything that plugs into it.

Back in 1651 the British philosopher Thomas Hobbes wrote

that life would be "nasty, brutish and short" if human beings didn't live in societies where we shared our innovations. And that was then. We've come a long way since. We should be so far away from a nasty, brutish existence that it's simply inconceivable to us. But what if a catastrophic event, like an orchestrated EMP attack, crippled our infrastructure? Do we have the knowledge to handle such an attack? Do we have the tools? Does someone out there have a plan? If a pulse shut down our electronic civilization how many aspects of our life would it impact? How long would the electricity stay off for? Would life become like Hobbes described? What's the worst case scenario?

# THE WORST CASE SCENARIO

If the power goes out, everything will change. This is how life after the pulse will look, in the moments, hours, days, and, yes, even months after. It's not a pretty picture.

To start with, there won't be any air raid sirens going off or any other such advance notice. That's not just because the few sirens that remain are poorly maintained and not connected to any national system. But because not even the people who'd sound the alarms would have any warning.

Back during the 13-day Cuban missile crisis in October 1962, Americans were on high alert the entire time. They'd practiced their drills. They knew where the closest bomb shelters were. It was, unfortunately, something they no doubt thought about constantly. And if Russia had launched a nuclear weapon, security personnel would have certainly known about it almost immediately. The public would have found out soon thereafter. This was the frame of mind many Americans were in - a state of fearful preparedness.

That's not the state we're in today. If the EMP attack comes from the KSM-3 - the North Korean satellite that hovered about the centre of North America right when the Metcalf transformer station was under siege - there'll be no notice at all. The U.S. uses what's called the Upgraded Early Warning Radar (UEWR) to detect incoming missile strikes. It's actually not one radar. It's an array of five radars with overlapping coverage. There's one on the east coast in Massachusetts, one on the west coast in California, one in Alaska, one in Greenland and one in North Yorkshire, England, that the British also employ to give them early warning of a ballistic missile test.

While the equipment is so modern that some parts are being updated at the time of this writing, they still don't have all of their bases covered. The locations were initially chosen and the first radars constructed during the Cold War, when the USSR was the key concern. The radars are well-positioned to detect launches from Russia. They should even do a decent job of keeping an eye on China. Once one of those missiles is up in the air, whether its purpose is a traditional nuclear detonation or an EMP, the United States will very likely notice it and have as much as thirty minutes to shoot it down.

There is one direction however that isn't well covered by these radars. It's to the south. At one point there's even a total blind spot. This at first doesn't seem like that big a problem. Not a single South American country has nuclear capabilities. This doesn't mean the southern exposure is no big deal though.

Just like any number of missiles, an EMP device fitted onto a missile can be detonated from vessels at sea. As we'll explore in a later chapter, EMP-related parts can crop up on ships passing through waters south of the United States. But the big worry about this blind spot to the south isn't that a launch from the ground - or water - will go undetected. It's that a satellite heading towards the U.S. on a northern trajectory launches an attack while passing through that blind spot. The KSM-3 could do this. It's already flies straight through the UEWR's southern weak point. If it happened, no one would see it coming and no one would be able to prove what exactly happened after the fact. It's the ultimate stealth attack.

When it comes to this attack, there won't be any warnings. There won't even be the whirring sound of a missile or bomb hurtling down through the sky as it nears its target. An electromagnetic pulse doesn't make a sound either. You can't hear it. You can't see it. You can't smell it. You can't touch it. It's almost as if it doesn't exist. And at first, for many, it'll seem like it doesn't.

The short term effects take place immediately. But not everyone notices them immediately. At least people who aren't using electronics won't. People out with the dog in the park, up on the roof replacing the shingles or studying by the window at a library won't notice any difference.

Even people walking about listening to music on their phones or texting a friend will just keep on doing it. There is not total certainty as to exactly which devices will be impacted and how they'll be impacted by a large scale attack. This is due to the sudden freeze in practical research that happened in the 1960s. What researchers have figured out is that there are three different electromagnetic pulse waveforms, that they label E1, E2 and E3. (*Blackout Wars* by Peter Pry, referenced in the bibliography, explains this in greater detail.)

A solar flare only produces the E3 waveform, which has a very long wavelength. Because of this it only comes into contact with long pieces of metal, such as power lines and telephone lines. This is the waveform that caused trouble to electronics during the Carrington Event and harmed the eastern grid during the Quebec ice storm. The way that other electronics are damaged by a solar storm is the E3 getting into these lines and then hitting anything connected to the grid. However man-made EMPs also generate E1 and E2 waveforms. The second one is similar to a lightning strike, which isn't the greatest of concerns.

The E1 wavelength is the big problem. This is a much

smaller wavelength that will work its way into anything over half a metre in length. This includes automobiles, appliances and any larger personal electronics. Personal cell phones won't go down, unless they're plugged in to recharge. Cell towers however will. You won't notice the grid has gone down while you're texting your friend. You'll just start to wonder what's happening when your message doesn't send or your friend doesn't text back.

As for cars and trucks, we tend to think of automobiles as off in a unique category, distinct from common electronics like televisions and major appliances. But these days the average vehicle relies on dozens of engine control units, which are essentially computer chips taking care of critical operations. And they're not just running the stereo system and the TV screens for the kids in the back to watch. Pull out those luxuries and there are still computer chips on board running basic operations.

After the passing of the U.S. Clean Air Act in 1963, North American automobile manufacturers installed different parts, electronic parts, to control their emissions. And by the 1980s various other electronics were put into cars. Now automobiles are full of computer parts, including kilometres of wiring. They're programmed with invisible but vital software. Cars in motion in the open air are prime targets for EMP wavelengths. The engines simply shut down. Only classic car enthusiasts and the few people who still drive decades old vehicles are in the clear. Their biggest challenge will be navigating around the gridlock of stalled cars blocking their way.

The key to understanding what is and isn't going to be hit by an EMP is figuring out if there's a metal conduit that connects the waveform to the object. Large metal items, like automobiles, become conduits themselves. So while cellular phones and many laptops are too small to be directly hit by any of the three waveforms, if they're in a car that gets hit they'll likely get zapped as well.

What about airplanes? Will an EMP make the thousands of planes that are above North America at any given time come crashing down? A launch will likely come from higher than the altitude at which plane flies. This means the pulse does hit the planes. Experts have called for more testing on this issue to fully understand its impacts. However there's little doubt the electronics are going to be hit in some way. Smaller planes with manual controls will be fine. Large commercial airplanes will have trouble. Pilots will have to manually land their planes. Odds are they won't be able to come down on the sorts of pristine airstrips they're used to. Hopefully they can navigate away from any forests or rock faces. The jury's still out on airplane effects, but it's safe to say you don't want to be up in the air without a parachute when an attack happens.

Even out in the country, life grinds to a stop. The cabin of today's farm tractors looks more like the command centre of a spaceship than a simple motor pulling a seeder along a field. Everything's done with the press of a button. And those commands are all executed by on board computers. Picture a farmer out in the field, with a tractor that suddenly goes dead. At first he'll think it's an isolated problem, until he goes into the house and realizes there's a larger problem afoot. As for people plugged into the grid when it happens – like office workers in town – they will know right away. They'll watch their office shut down right before their eyes. The lights turn off. The hum of the refrigerator goes quiet.

People slowly realize there's a problem. The first thing they do is attempt to contact family and friends. That gets them nowhere. Then they try to find out what "they" are saying. The ubiquitous "they" everyone is always talking about. The people out there who always know how this complicated world works when the rest of us are focused on our own lives. The people who always have a handle on the situation. There's got to be a "they" when it comes to something like this, right? Wrong.

People look to social media on their phone or laptop first.

Nothing. They try to turn on the television. It doesn't work. They turn on the old radio they've got in the closet that they haven't gotten away to throwing out yet. Finally. Something. The little device works. But it's only static. Move the dial this way and that way. Most of the stations aren't on. Leave it on. Keeping turning the dial. Maybe some far away station will broadcast something soon.

What do we do now? For a population so used to being plugged in, connected, always in the know, it's a strange feeling. Go sit on the couch. Read a book. Wait a couple hours until the power comes back on. But it's hard to do. Something's not quite right. This feels different than those times back when you were a kid and the power goes out so your parents bring out the flash lights and tell ghost stories until the lights come back on. After all, what's with the stalled cars? Maybe there is a ghost. Well not actually a ghost. But something unseen that's eerily wrong. Something nefarious in the works. Maybe the lights won't come back on.

People in buildings with emergency lighting feel more secure. The presence of some sort of electricity is a sign that all's not lost, that there's still something greater than them running the underbelly of society. They're not alone. But emergency lighting and exit signs are often fuelled by battery power. They've only got so many hours of juice in them. Their main purpose is to help people navigate safely out of buildings.

For others, the lights around them only go off for several seconds. Then they come back on. Emergency generators have kicked in. It's a reassuring feeling. In most jurisdictions it's the law for facilities such as hospitals to have emergency power systems. And the law even determines how many seconds the power can be off. It's usually around ten. But emergency generators are just that. For emergencies. They don't last forever. They're powered by diesel fuel and the tanks aren't that large. Many generators only have sufficient fuel to remain on for 24 to 72 hours. That's usually enough time for the power to come back on or, if necessary, for a truck to deliver more diesel.

Aside from these minor hiccups, in the first few hours all is well. We still don't know exactly what happened. But sit tight and life will soon return to normal. Hopefully. Meanwhile, "they" aren't any better informed and they're not working on a solution. They are just as confused. The utility companies are rushing about their head offices and generating stations trying to figure out the problem. The authorities in intelligence and national defence probably have some clue about what happened, but they can't be certain. The key people in government are waiting to hear from experts on their options. But, as we'll learn in later chapters, for too many government departments there are no options. There is no plan. "They" are in the dark.

The vast majority of people without power get by for the first few hours just fine. It's something different. Maybe even a little fun. For many people the main worry is that if the power doesn't come back on soon they'll need to throw out some of the food thawing in the freezer. The dangers will soon increase though. With every passing hour, the risks that lie ahead loom larger.

On August 14, 2003 a utility company in Ohio learned the risks of their shoddy maintenance work by letting trees near their power lines grow too big. The lines sagged and struck the trees. It sounds like a minor problem. But that sloppy housekeeping resulted in a cascading failure that took down the power for much of the north-east region, resulting in one of the worst power outages in history. New York City, Toronto, Detroit and other cities went dark. Fifty million people went without power. For many people, the lights went back on that evening, only hours later. For others, it took a couple days.

Yet despite the relative brevity of the black out, people still died. In Toronto, a burn victim recovering from a previous electrical accident died because the air conditioning he relied on to keep his skin grafts cool wasn't on. In Detroit and New York City, people died from carbon monoxide poisoning. In one Michigan town a man died from a fire caused by burning candles. Fire departments responded to thousands of calls about minor fires caused by candles. Thankfully the death count wasn't higher.

This is where the effects of an EMP attack diverge from those of a regular blackout. The short term death count, tragically, will be higher. A lot of fire trucks – which are basically giant computers, just like most other automobiles – won't function. Trucks sitting outside of the station bays exposed to the wavelengths will be down. Fewer trucks on the road will lead to significantly longer responses times, if they're even able to receive communications about emergencies in the first place. Minor fires that are normally easy to tackle have time to expand and become harder to knock down. There's the added challenge that water pressure is also reliant on electricity. You can't get water to flow against gravity naturally. At least not without a hand pump. If you're pumping it up from the ground, some sort of machine is involved. The firefighters on site might not even be able to use all of their equipment or get water flowing from fire hydrants on the streets and standpipes connected to large buildings. Besides, if the telecommunications grid goes down, good luck getting ahold of 9-1-1 to report any fires. Or anything else.

Sunnybrook Hospital, Canada's largest single-site hospital, went without power for 39 hours during a Toronto black-out caused by an ice storm one day in December 2013. The power went off at 4 a.m. While the generators initially turned on, at one point all of them except for one shut off for two hours. Their neonatal intensive care unit, which cares for premature babies who are connected to life support machines, lost all power. Six babies were transferred to another hospital. A number of other patient services were cancelled during this period. A world renowned hospital in a developed country that routinely receives donations from the richest people in the country couldn't care for babies because its generators failed following a storm. Something's not right here. (The hospital understands this and is now in the process of upgrading its generators.)

This was during a simple storm. How would Sunnybrook have fared after an EMP attack? At one point during the ice storm, they brought in an additional off-site generator. After an EMP, if you can even identify a functioning one good luck getting it delivered. Neither will tanker trucks, shut down by the wavelengths, be on standby to deliver top-up fuel for the generators. The high-tech ambulance equipment the fragile babies relied on for their hospital transfer would also be useless. And who's to say the hospital they're being transferred to isn't suffering from its own problems? There would be no other secure hospital to accept them.

The bottom line is that within days after an EMP attack, most life support machines shut down. This doesn't just impact hospital patients. Anyone in a retirement home needing support or using a device at home – they're all at risk. People will die. Newborn babies will die. Some people will be kept alive by around the clock staff and loved ones operating devices like manual ventilators. But there'll soon come a time when hospital staff will make hard choices about where to allocate their stretched resources.

As if these challenges aren't enough, they're just what arises in the short-term, within hours and days. The mid-term effects start to cause trouble as the days turn into weeks. The biggest challenge in this period is securing food and water. This is when the concept of the electronic civilization really hits home.

In 1851, Canada's population was 87% rural. The food distribution system was much more local than it is today. People grew their own food. Or got it from a nearby location. The food distribution network wasn't all that complicated. But things quickly changed. In 1941, there were 1 million male workers employed in agriculture. Thirty years later that number dropped to 400,000. While the number of farm workers dropped in half, the general population doubled. Fewer people made food for an increasing population. The math somehow works against the odds. Up to a point. Clearly people are still getting enough calories per day and then some, thanks to automation and centralization. But these industry changes aren't all for the better. Fewer people have the knowledge to maintain the food chain at the same time it's becoming increasingly complex and physically removed from the people it serves. Now, the numbers are reversed. Today Canada is over 80% urban.

Grocery stores and corner shops in towns and cities generally only have several days worth of food on hand. Regional hubs and storage facilities have about a month's worth of supplies. It's cold comfort. Even those 30 days worth of sustenance are only helpful to people who live nearby. If many of the trucks don't work, the distribution chain fails. And it's not like food hubs have a better grade of generators and a better fuel supply than hospitals. They can't save what food they have from spoiling in the event of prolonged a blackout.

As technology dug deep roots into farming, it too became a fragile part of the electronic civilization. Today most farming has an electrical component. Following a pulse, tractors sit idle, machines that tend to livestock shut down. This means store shelves go bare. People in the cities go hungry, while perishable goods rot in food terminals.

The water system's the same. Like with the Maroochy water incident, these systems can go haywire. Valves open at the wrong time or let the wrong volume flow. Sewers back up. Then, as the generators fail, the pumps just stop working. A basic necessity of life that 21st century people have come to believe is instantly available for them when they turn on the tap suddenly splutters to a stop.

It's happened before. The havoc caused by Hurricane Katrina in 2005 offers a small-scale version of what post-

EMP life could very well degenerate into. As the White House's official report on Katrina's impacts on New Orleans and surrounding areas notes: "Nearly a quarter of a million people in shelters relied on shipments of ice, food, and water to meet their basic needs."

Modern hunger and food shortages are rarely caused by insufficient amounts of food. Studies have proven they're more often caused by poor distribution of food. We've created a society where we can no longer look after our own basic needs if the grid goes down.

Katrina resulted in over 1000 deaths. Over 90% of the population of New Orleans was evacuated. The economic and recovery costs were over $110 billion. Yet this was only one region of the United States. The rest of the country was on hand to lend support – even if the speed and quality of that support garnered harsh criticism during and after the fact. Other states and cities sent their workers down to help. Non-governmental relief agencies got to work. Regular people sent donations of money or goods. It's not like the whole country was in the dark. North America can mobilize its resources when one area is devastated. But everyone is affected by a wide scale EMP. Everyone is on their own. Help will not be on its way because whoever could help is looking after their own families.

For those healthy people with access to basic resources who survive the short and medium-term effects of the attack, they're not over the worst of it yet. They'll have to contend with the long-term ones. Urban-dwellers will flee the cities in search of food and safety. Rural residents will try to hunt, scavenge and grow their own food. As civil society breaks down, people will form alliances, packs and gangs. There will be fights over food, water and shelter. Everything we took for granted about the comforts and safety of modern living will disappear. Within weeks after the attack, people will start to live like the characters in post-apocalyptic shows and movies that they were watching for simple entertainment right before the attack hit.

There are already people in North America gearing up for such a societal breakdown. A Chicago Tribune feature from August 2016 profiled the experiences of one group of "preppers" or survivalists: "Don and Jonna Bradway recently cashed out of the stock market and invested in gold and silver. They have stockpiled food and ammunition in the event of a total economic collapse or some other calamity commonly known around here as 'The End of the World As We Know It' or 'SHTF' - the day something hits the fan," the story begins. "They melt lead to make their own bullets for sport shooting and hunting - or to defend themselves against marauders in a world-ending cataclysm." The Bradways are more on the prepared side than most North Americans, to put it lightly. They're far from alone. Estimates put the number of North Americans living some form of the preppers lifestyle in the low hundreds of thousands.

It's hard to imagine how it could ever come to the point where civil society breaks down as badly as this and people are left to fend for themselves. Is it actually possible that the grid goes down for such a long period of time that these medium and long-term effects actually come to pass? After all, North America's infrastructure grid is highly regulated. It has layers upon layers of emergency protocols built into it. There are many thousands of knowledgeable, educated and well-paid engineers and technicians who understand how the grid functions and, therefore, how to bring it back online. The post-apocalyptic scenario the preppers await just shouldn't be possible. But it is. Because there are a couple of key differences between what happens when the grid goes down after an EMP attack compared to a regular blackout.

The big difference is that electronic circuits don't just turn off after an attack, like they do during a black out. They're damaged. Or totally destroyed. During a regular black out, the way the affected area gets back online is through one of two methods. Power is brought in from external parts of the transmission network. It can take time as it needs to be done slowly and cautiously. But piece by piece, power stations turn back on. The other option is a black start. This is when a smaller diesel generator is used to fire up a slightly larger generator that's connected to the grid, which in turn is used to start an even larger one and so on.

If one transmission station or power generator or set of lines in any given area is destroyed then power can simply be re-routed. But if everything within an area is taken out, the problem becomes much larger. All kinds of parts need to be replaced before progress can be made. And it's not just as simple as having a worker show up with a tool kit and switch out a couple of circuits. Not all of these parties will just be readily accessible when needed in sufficient quantities.

While electricity comes from power plants, it doesn't go directly from those plants into homes and businesses and elsewhere. It first has to stop off at a middle-man, which is usually an extra-high voltage (EHV) transformer. These are the large box-like structures found throughout towns and cities that are housed behind chain-link fences, usually in some out of the way place like beside a park or field. Some are larger than others. Sometimes there are only a couple of them placed together. Other times there are dozens at one site. The Metcalf transformer station is an example of a larger collection of EHVs. If these are damaged, then power can't get from its source to the point of consumption.

"Failure of a single unit could result in temporary service interruption and considerable revenue loss, as well as incur replacement and other collateral costs," notes a 2012 report from the U.S. Department of Energy. The very next sentence presents the real problem though: "Should several of these units fail catastrophically, it will be challenging to replace them."

That same year the Science and Technology division of the Department of Homeland Security also put its focus on the issue. "EHV transformers are huge, weighing hundreds of tons, making them difficult to transport – in some cases specialized rail cars must be used (and there is a limited supply of these)," their website explains. "Many of the EHV transformers installed in the U.S. are approaching or exceeding the end of their design lifetimes (approx 30-40 years), increasing their vulnerability to failure. Although the industry does maintain limited spares, the ability to quickly and rapidly replace several transformers at once would still be a challenge."

It's a recipe for disaster. And the government admits it's a type of disaster that they're far from immune to: "Many events could cause such catastrophic failures: terrorism of course, but also such unavoidable events as natural disasters: hurricanes and tornadoes, and even solar plasma flares [such as the Carrington Event] disrupting Earth's magnetic field."

If a number of EHVs go down at the same time, there's going to be a major problem. There are approximately 2000 EHVs in America alone. Yet there are only four manufacturing facilities in the U.S. and two in Canada that make them. None of them are on the West coast. And none of them are close to major urban centres like New York City or Toronto. The average time to make one ranges from six months to a year. That's the minimum a region will be out of power for if their EHVs are taken out of commission and if they don't have an extra one randomly waiting nearby, which they most certainly do not. And that's assuming that these EHV plants themselves haven't been impacted by any attack, which they certainly could be. It's not like they can continue building their products if the power grid around them is shut off. It takes power to make power.

The American and Canadian plants aren't the only ones in business though. New ones can always be ordered from abroad. In fact, right now 85% of America's power transformers are actually imported. South Korea is a dominant player in the market and China has 30 plants that make EHVs. But the delivery time is significantly longer. And if the grid is down, how exactly will the EHVs be transported to their destination? Once the parts are delivered by boat to a port, they still need to make their way inland. Yet transportation will have been seriously affected by the attack and one of the whole reasons the new EHVs will be needed is to help restart those very transportation networks to deliver much-needed supplies around the country.

One of the key challenges around rebuilding the electronic civilization after an attack is this chicken-and-egg scenario. What comes first? You need power and electricity to do many of the things required to turn the power and electricity back on.

In one study conducted by a consulting firm into the economic impacts of an attack, the authors spoke with various experts to get their take on just how many EHVs would be damaged. They give the low range at a comfortable 10% of broken transformers. They also offer a low range estimate of getting a replacement up and running in two and a half months. Using these numbers, it's a scenario emergency management personnel could work around. If only a tenth of EHVs are down, surely the power from the remaining grid can be rerouted while they wait for replacements.

But the high range estimates are truly troublesome. A full 70% of EHVs being knocked offline, the study points out. It'll be quite a challenge to make up for that lost capacity. Entire sections of the country could be locked off from power. And the high-end estimate replacement times for transformers are 33 months. That's almost three years to get a new EHV. There's also no telling how the damage to actual power plants will overlap with the damage to the transformers distributing that power. The surviving EHVs are only useful if the generating stations near them are functioning.

The Department of Homeland Security was rightfully concerned about these harrowing facts. Tornados, cyber-warfare and other aggressors can also take down EHVs. So DHS wanted to see if there was a way to replace a transformer much faster. In 2012 they embarked on a project known as RecX – short for Recovery Transformer. They teamed up with a private utility company in Houston, Texas – CenterPoint Energy, which would be on the receiving end of the new EHV. A transformer was made 750 miles away in St. Louis, Missouri. It was only 60 tons, compared to the usual transformer weight of hundreds of tons. They designed it to be much faster to transport and install. Once it was made and sitting on the shelf in St. Louis, a convoy of trucks took it to its destination and got it up and running in less than a week. The RecX was celebrated as a major reduction in EHV replacement time, a sign that the electronic civilization could be rebooted quicker if needed. It was a success – as far as pilot projects go. But that's all it remains. A test. It's not the standard and it's not like every utility company and EHV manufacturer knows how to do it this way. Besides, the whole project actually took over three years from conception to completion. Sure, it was smaller and could be shipped faster. However that's about it.

"Ninety percent of consumer power passes through these critical pieces of equipment at some point on the transmission grid," a 2014 report on the RecX by DHS explains. "If these transformers fail – especially in large numbers – the nation could face a major, potentially long term, blackout." Despite the success of the RecX experiment, these inconvenient truths that inspired the project still hold true today.

The technical details detract from the big picture of the worst case scenario. The bottom line is this: If the power goes out for many months, even stretching into years, life changes. If, say, North Korea attaches a nuclear warhead onto their KSM-3 satellite and detonates it as it passes over the Midwest, the electronic civilization halts. If a terrorist group manages to sneak a long range ballistic missile onto a freighter that gets near the east coast and then launches it, our entire way of life halts. Water. Food. Transportation.

Communication. It all changes. We revert to life as it was many decades ago. However we don't have the knowledge or tools to live that way. And that era wasn't able to sustain today's population. Canada now has 35 million people. In 1950, the number was 13 million. In 1900, it was 3 million. The United States has a population of 320 million. In 1950, it was 150 million. In 1900, it was 75 million. This is the worst case scenario of an EMP attack: The majority of North Americans die in a short period of time. It's that simple.

# THE ARMING OF THE CHONG CHON GANG

As members of the Panama navy stormed a cargo vessel they'd forced ashore, they expected to find a hidden stash of illicit substances. It was July 2013 and they were on the lookout for drugs after receiving a tip from intelligence sources. But what they found was much more troubling.

The Chong Chon Gang looks like a typical cargo vessel. It has four floors of living quarters with small windows. Its white walls are stained by sea water. Dozens of vessels pass through the Panama Canal every day and many of them look just like this one. Aside from one detail: the large North Korean flag at the top.

The storming of the ship wasn't a typical boarding either. Panamanian inspection agents are used to looking for drugs. That's their specialty and it's what they were doing that day. Both Interpol and the U.S. State Department regularly warn about the high level of drug trafficking that happens via the Canal. While many ships trafficking drugs make it through, others do not. The crew of these vessels are no doubt nervous whenever passing through a policed area. But the North Korean crew of the Chong Chon Gang was particularly frantic after their ship was boarded. The captain tried to commit suicide and the crew began to riot. This wasn't going to be the usual drug raid.

Once officers made their way through the vessel, they found thousands of small bags marked "Cuban Raw Sugar" down in the hull. This wasn't a rare shipment to be carrying. The Chong Chon Gang was making its way home from Cuba and was stopped near the Atlantic side of the canal in Colon. None of this is out of the ordinary. Raw sugar is Cuba's main export and it heads mostly to China, the Netherlands, Senegal and the United Kingdom. It wouldn't raise any red flags if the Democratic People's Republic of Korea ordered a shipment of the sweet stuff.

Now even if there was more than just sugar in the hull of the vessel, this wouldn't have been a big surprise either way. It's not uncommon for drugs to be tucked away inside sugar shipments. Just like in scenes from the movies, smugglers dress up drugs as all sorts of otherwise legal shipments. Sugar would be an obvious disguise for a substance like cocaine. Yet when officials sorted through the bags of sugar and brought out x-ray scanners to see in them and through them, it wasn't drugs they discovered. The Chong Chon Gang was transporting weapons. And lots of them. The hidden goods included two MiG supersonic jet fighters, nine missiles and two anti-aircraft missile devices.

Cuba's foreign ministry said the materials were just heading up to North Korea to be repaired. Then they'd come back. Besides, they were disassembled. Plus, they were pretty much obsolete. No big worry. That was the way Cuba tried to sell the story.

Was it the full story? And if it's true, does it make the whole incident less unsettling? After all, the ship could have made it through without incident. Just like drug shipments make it through the Panama Canal - a corridor to America - on a regular basis. The United States and its allies monitor North Korean vessels in transit. To their best of their ability. And to the best of their ability they get a sense of what's on board. Yet intelligence told them it was drugs and it wasn't. Maybe next time there's a shipment like this, intel will completely slip up and even say the ship is fine, there's nothing bad on it, look the other way. And then a repeat of the Chong Chon Gang incident happens. Except this time the contraband weapons make it through successfully without anyone being any the wiser.

This story matters even more when you keep in mind North Korea's nuclear program is picking up steam. Their first nuclear test launch was in 2006 and the yield of the detonation was less than a kiloton. Since then, they've made strides. In September 2016 they conducted their fifth detonation. It was 10 kilotons. External sources confirmed seismic activity readings were consistent with the sort of underground detonation the rogue state claimed to have pulled off.

It gets worse. On the 1st of January 2017, Kim Jong Un gave a New Year's address announcing what many analysts had been worrying about for years. The dictator boasted they were on the verge of test-launching an intercontinental ballistic missile (ICBM). As the name suggests, that's the type of powerful missile you need to lob a nuclear weapon all the way from one continent to another. They're the sort of weapons the USSR and America would have used to exchange nuclear attacks during the Cold War. "The ICBM will be launched anytime and anywhere determined by the surprise headquarters of the DPRK," a government spokesperson announced.

Around the time this book went to print, President Donald Trump was getting serious about confronting North Korea over their nuclear aspirations. On the weekend of April 8, 2017, Trump was finishing a series of meetings with Chinese President Xi Jinping that included talk of containing North Korea. Soon after those meetings ended, a U.S. Navy strike group made its way up the Korean Peninsula to guard as a deterrent against the north's ambitions. This failed to calm the situation. North Korea responded by saying they were "ready for war" if need be. Trump fired back on social media that "North Korea is looking for trouble. If China decides to help, that would be great. If not, we will solve the problem without them!"

So far North Korea's tests are largely viewed as just that. Tests. Sure, it's cold comfort for a close neighbour like South Korea. But tests alone don't exactly turn the global power dynamic upside down. Although what happens when they do get ICBMs functioning? The yield of nuclear weapons they've been testing remains low. The detonations would be minor compared to the types of bombs the U.S. can fire back at them. That's the underlying concept of mutual deterrence and mutual destruction. It's the idea that no country, hopefully not even North Korea on its nuttier days, will fire a nuclear weapon at another nuclear armed country because once it's up in the air the other country fires theirs back. There's nothing gained. Just a lot of hurt caused on both sides. All of that calculus changes though when it comes to setting off an electromagnetic pulse. If they place an EMP in the air through more clandestine means, like dropping it from a satellite or launching from a vessel tucked away on the eastern seaboard, they might be able to detonate it. Once that detonation happens, the grid will falter or fail. Good luck firing back. There's no mutual assured destruction for that category. Only a North America in the dark. The added bad news is the small yields Kim Jong Un's country has detonated to date are sufficient to generate an EMP.

Among the questionable goods stowed aboard the Chong Chon Gang was an SA-2. That's a high-altitude air defence missile launcher. It doesn't boast much horizontal range compared to an ICBM. It only travels about 50 kilometres. But it's the vertical reach that's of note. It can go up over 20 kilometres. Detonate a warhead up that high in the sky and there's a good chance the energy would interact with gamma rays and send down an EMP. Maybe the Chong Chon Gang was delivering these parts to a vessel planning to launch it. It's even possible that, had a few variables been different, the Chong Chon Gang itself could have set off an attack. And if it made it to a different spot, had it crept up the Atlantic, it could have set one off that took down the eastern grid, knocking New York City, Washington, Toronto and Ottawa back into a pre-industrial era lifestyle until the states out west rescued them. Remember, an EMP burst doesn't have to happen directly over a region to impact it. Everything in the sightline of the detonation – from many kilometres up in the sky – will be hit.

This is a key point to keep in mind. An atmospheric launch from a satellite like the KSM-3 that detonates right in the middle of the continent isn't the only way to unleash an EMP attack on North America. It's without a doubt the best. It's certainly the one that'll bring about the worst case scenario or effects close enough to it. But it's far from the only option enemy actors have to cause catastrophic damage. There are a number of countries and even non-state actors aware of and at work on EMP or EMP-related technologies.

There's already been one massive man-made EMP attack in human history. It's known as Test 184. And it took down the entire grid of the ninth largest country in the world. Few people talk about it and even less know about it. This is because it all took place behind the Iron Curtain.

The Central Asian country of Kazakhstan has been tangled up with Russia for centuries. Russia encroached on their lands starting in the 1700s. Later, they were de facto absorbed into the Russian Empire. Then they became part of the Soviet Union in 1936. During this time, the USSR did some of its dirty work in Kazakhstan. Near the Kazakh steppes – a large expanse of open grassland – they set up some of the notorious Gulag work camps where many people died, mostly from famine. It's where the Soviets also constructed the nuclear test facility Semipalatinsk-21.

Back in 1962, the Americans almost stumbled upon the truth about high-altitude EMP effects during the Starfish Prime tests. It was only after seeing it play out on their radar equipment and during the Hawaiian street light incident that they strove to replicate the effects. But they couldn't organize enough above ground tests before the partial test-ban treaty was enacted. Then there was also the fact that the Americans just weren't willing to launch tests on large scale, live environments. Even without the test ban, they'd never truly experiment with a pulse to learn how it impacted real life communities. So EMP was always destined to remain something of a theoretical field for American science and military research.

That rule didn't apply for the Soviets though. They had no qualms conducting a live test. And it was in an area in the middle of Kazakhstan where they decided to spring Test 184 on an unsuspecting population. While Semipalatinsk-21 conducted hundreds of tests over the years from Kazakhstan, it wasn't the only such facility in the USSR. There was also Kapustin Yar, one of the few publicly known test centres. The facility – which is still in operation today – is in the southwest of the country, with Moscow to the north, Ukraine to the west and Kazakhstan just a hundred kilometres or so to the east.

It's the perfect location. The ideal spot from which to fire a missile into and above the centre of Kazakhstan. That's what K Project, as the EMP tests were dubbed, aimed to do.

The city of Zhezqazghan, with a current population of 90,000, was Test 184's target. From there, it's a 2,800 kilometre car trip to the Kapustin Yar launch site. This was no short range missile launch. They're the same distance apart as Moscow is from London, England. In fairness, they didn't aim for it to directly hit the city. And they may not have known precisely how far the effects would be felt. Their actual target was simply the coordinates 47.78 N, 65.329 E. That's about 200 kilometres to the west of the city, in the middle of nowhere.

And on October 22, 1962 – three months after the Starfish Prime launch – the USSR fired Test 184 towards those coordinates. It detonated almost 300 kilometres in the atmosphere above its target.

The effects of the detonation were widespread. They weren't just constrained to places that could be described as "in the middle nowhere". It impacted in various ways a sizeable number of the country's population, which at the time was just under 10 million people. As far as 1000 kilometres away, radio devices malfunctioned. An underground power line 1000 kilometres long that connected major cities failed. Overhead power lines burnt out and even fell to the ground. The fuses in telephone lines blew. Military and civilian electronics were destroyed, damaged or failed. And, most critically of all, power generators shut down or were damaged. A power station in Karaganda, one of the country's largest cities, was even set ablaze.

If these effects aren't bad enough, think about what would happen if Test 184 hit Kazakhstan today. Things would be much worse. Many of the electronics destroyed back then were sturdy devices. They contained parts like vacuum tubes, the sort of pieces that were in the behemoth ENIAC computer. Plus, it was a powerful launch. Its energy readings were twice as strong as the ones recorded during the 1989 Quebec power storm that temporarily shut down the grid.

EMP remains a largely unexplored subject that politicians, the media and public know little if anything about. But we know even less about Test 184. It's one of the Soviet era's better secrets. There are no noteworthy studies of it conducted by Western academics and researchers. There are no memoirs released by those who suffered from it. There's been no official apology from Russia for their actions. The West only knows about it from the few occasions that Russian scientists briefly peeled back the curtain on the phenomenon. Like how General Vladimir Loberev spoke about it in 1994 in France at the 1994 EUROEM Conference, the annual European Electromagnetics Symposium that still happens today. Or how a Russian scientist discussed it during a meeting with the Lawrence Livermore National

Laboratory, a government-funded research facility housed at the University of California.

There are still holes in the story. Eerie ones. For instance, we know K Project involved a number of tests. But Test 184 is the only one we have extensive information about. Is this because the rest failed? Is it because they were less impressive? Or is it because there were others that did succeed and were even more damaging? We also don't know about the recovery process. Was it easy? Could they reboot the system quickly enough? Or were transformers out for many months? Was there societal collapse? Did Kazakhs receive sufficient food and water following the test? Did people die in the aftermath?

That's what we don't know. And what we do know is clearly scant. But we know for certain that Russia successfully tested an EMP detonation on an inhabited country – a member of their own union no less. They fired it from almost 3000 kilometres away. From that distance, they could easily launch a missile from international waters into another country. Even deep into another country. Thankfully Russia is far more stable than North Korea and better disposed towards the United States and Canada than the rogue state. But none of that changes the fact they have the technology, they know how to use it and they've done it before.

It's unknown exactly how much of an interest the Russians currently have in EMP. We do know it hasn't completely slipped off their list though. It's still very much a part of the conversations they have about strategy and military doctrine. Major General Vladimir Slipchenko was one of Russia's leading military theorists before he died in 2005. He coined the concept "sixth generation warfare". The previous generations of warfare, which are a classification acknowledged by military experts around the world, show the evolution from soldiers standing in lines in the middle of fields to the trench warfare of the first World War to modern nuclear weaponry. But Slipchenko dubbed the newest generation "no contact wars". The sixth generation is about hitting an enemy combatant from all possible avenues, mostly the ones that don't involve physical contact. Prime examples are cyber sabotage and EMP attacks. Strong nations can be brought to their knees by a well coordinated no contact war.

It's usually difficult to unearth evidence of military and government figures having these conversations. They usually happen behind closed doors at places like the Russian Academy of Military Sciences. They're not all that forthright about what they're doing. It's not like they send out weekly press releases to Western media. However English readers did get a unique glimpse into the long-term strategic thinking of one leading Russian expert in a 1998 book released by the major American publisher Routledge. *If War Comes Tomorrow? The Contours of Future Armed Conflict* is its title, written by General Makhmut Gareev. The 93-year-old Second World War veteran is a former deputy chief of the USSR's armed forces and currently the head of the military sciences academy.

In the English translation of his book, which is something of a military textbook in Russia, he makes it clear that EMP is to play a role in the future of non-traditional warfare: "Further perfection of nuclear weapons is aimed at increased accuracy against military objects, the ability of nuclear missiles to overcome systems of antimissile defense, the creation of third-generation nuclear ammunitions with reduced radioactive emission, super-EMP (electro-magnetic pulse), which employs an electromagnetic pulse of high voltage capable of putting out of action all electronic systems of the enemy, and which affects ground-based missile complexes, radar-tracking stations, communication facilities, methods of automation, and counter-electronic warfare." It's clear EMP is still a topic at play in Russian military thinking. Even if it's just on the theoretical side. And this in itself is a problem. It's not like they've disavowed its usage, even though they know first-hand its catastrophic effects. It's not like they were horrified by Test 184 and vowed to forever turn the page on the subject. It's still in the game.

The good news is that Russia isn't a fringe rogue state. It's a massive country that predominantly views nuclear weapons as a deterrent and not an offensive tool, just like the United States does. No matter what a person's concerns are about President Vladimir Putin's modus operandi and his incursions into Crimea and operations in Ukraine, there's a certain level of hubris that a country of that size just doesn't engage in. The same can't be said for how others view offensive strikes. Like rogue states. Or terrorist organizations. And while Russia may be less inclined to launch an unprovoked strike themselves, that doesn't mean they're not willing to assist far less stable actors in acquiring the technology.

In 2013 South Korea's National Intelligence Service revealed in a report to their parliament that Russia had actually sold EMP weapons to North Korea. The major wire service Agence France-Presse reported that Kim Jong-un aims to develop his own versions of the weapons. Little is known now about the pace and success of the program. But if it's anything like their nuclear detonations program, they will have made strides in the past few years.

South Korea believes their neighbour to the north is planning to use the weapons to attack them and disable their military equipment. It could very well be true. However the South's assessment wasn't an international one. It was solely focused on reporting to their government about threats to their homeland. North Korea could just as well be planning to strengthen their arsenal for a run at another country as well. Or to arm a vessel such as the Chong Chon Gang.

This is why Russia's expertise in EMP weaponry is so worrisome. They sell weapons to many troubled regimes that in turn sell weapons to even less stable actors.

There's Iran, which supplies weapons to Hamas and Hezbollah. There's Pakistan, which is considered by many politicians, academics and public intellectuals to be a state sponsor of terrorism. There's Saudi Arabia, from which the majority of 9/11 hijackers originated and which provides funding for many of the most ideologically troubling madrassas – Islamist schools – around the world. Then there's Syria, which is a hotbed for violent sectarian groups right now after the destabilization caused by civil war and the rise of the Islamic State. Any of these groups, including ISIS itself, could get their hands on these parts.

The Russian military is, based on the above evidence, a major source of EMP innovation and proliferation in the world today. This has the potential to make the world less stable. North Korea, which received parts from Russia, has made it abundantly clear they want to see both the United States and South Korea turned into a "heap of ashes".

As for Iran, in September 2016, a senior Iranian commander said they've got the ability to "raze the Zionist regime in less than eight minutes." The Zionist regime in question is of course Israel. Hostility towards Israel is nothing new. It reached a peak during President Mahmoud Ahmadinejad's time in office. He called for the annihilation of Israel multiple times. While he retired from politics in 2013, it's clear from the commander's remarks that the sentiment remains among the upper brass. It's an understatement to say Iran is still hostile to Israel and it's more than possible EMP weapons are in their arsenal. And if they're not, they're definitely within reach.

But even if the countries Russia deals with don't act on their intentions, they in turn have relationships with groups that do. Playing "follow the pieces" is a hard game when it comes to tracking EMP-related parts. When states are fragile, they're easier to steal from and manipulate. The various terrorist and guerrilla groups operating in Syria could get their hands on the state's various weapons, parts and plans. In the case of Pakistan, they may hand over such weapons willingly. That's the definition of being a state sponsor of terrorism.

One can explore and debate the finer details over whether the countries Russia has sold weapons to pose direct existential threats to Western nations. But there's no denying these countries are home to terrorists who aim to pose such threats. These are the same sorts of terrorists who saw nothing wrong with hijacking passenger planes and crashing them into buildings. It's unlikely the sort of Al-Qaeda adherents behind 9/11 would get squeamish at the prospect of detonating a device that doesn't actually even kill anyone immediately.

No terrorist group has yet to successfully detonate an EMP device on a large scale. But the world's thugs clearly know the advantages of taking down power grids. They've done it before. The Knights Templar drug cartel is one of the dominant gangs in the Mexican state of Michoacán. There is little savagery they won't engage in. From street shoot-outs to beheadings to testing their members' loyalty by making them eat human hearts, they've done it all. So to their individual members, firing machine guns into transformer stations was likely one of their lighter assignments.

This is what they did on October 27, 2013. It was a coordinated attack. They fired guns and threw bombs and Molotov cocktails at electrical stations across the Michoacán region. They blew up at least nine of them. This took down the grid for a dozen towns and cities, including much of Morelia, the state capital. A million people were without power. Then they torched four gasoline stations for added effect. A dozen people died in the ensuing battles. The lights remained off into the following day.

The Knights Templar succeeded in taking down the grid precisely where they wanted it down. The organized and hierarchical gang could have gone after a wider net of transformers if they'd wanted. Maybe if they get wind of an EMP device on the international black market, they might just put in an order.

It wouldn't be unprecedented. Terrorists once put the entire nation of Yemen at their mercy. The Knights Templar assault was nothing compared to what al-Qaeda in the Arabian Peninsula launched the following year in Yemen. They did something that had never been done before. On June 9, 2014, they plunged an entire country into darkness. The second largest country in the Arabian Peninsula. All 24 million people went without power for at least two days. It was longer in some parts of the country.

Small and brief power outages are tragically common in the already impoverished and fractured country. Warlords and power brokers shut off the electricity in regions to serve as leverage for their demands. This causes fuel shortages and other headaches that make the already challenging daily struggle in Yemen even worse. But never before had the entire grid gone down, and not for longer than a day. "Technical teams were attempting to fix the lines after the first assault," Al Jazeera reported at the time, "but gunmen sabotaged them a second time and prevented technicians from fixing them again."

Despite their derivative name, AQAP is no al-Qaeda copy cat. They're now the main deal. Following the death of Osama bin Laden, the Yemen branch is considered the most dangerous and powerful branch of al-Qaeda operating in the world. AQAP aren't isolated to their home turf either. They were responsible for the Charlie Hebdo killings on January 7, 2015, in Paris. If they want to strike abroad, they do. Gangs like the Knights Templar know the value of knocking out the power grid. Leading terrorist groups know the value of it. Don't think they wouldn't want to keep an EMP device on hand.

At first it seems like a stretch to imagine groups like these having the know-how and capacity to acquire, store and maintain an EMP weapon. But labeling them a gang, even if that's what they are, gives a false impression of their abilities. It suggests they're a rag-tag, fly-by-night operation. Likewise calling groups like AQAP or ISIS mere terror or extremist groups downplays their might. These labels evoke images of street corner gangsters conducting their business in back alleys or roving terror cells camped out in the bushes. But the truth is the complete opposite.

AQAP and ISIS are, at least at the peak of their powers, robust organizations. It's almost more appropriate to use terms like institutions, corporations or governments to describe them - even if they don't satisfy all of the legal or technical qualifications. These groups have organizational flowcharts. Offices. Sometimes even entire office buildings. Uniforms. Bureaucrats. Accountants. Filing cabinets full of paperwork. They follow best practices, in their own bizarre way. They have internal codes of conduct, even if it's not the type of conduct most reasonable human beings aspire to follow.

At the height of its power, ISIS was in control of civic institutions and infrastructure. They managed water filtration plants, generating stations, oil refineries, dams and more. It's folly to underestimate the technical and operational competence of gangsters and terrorists. Taken one by one, none of these threads presents a direct cause for major alarm. The arming of the Chong Chon Gang. The havoc caused by Test 184. The targeted assaults on the Metcalf transformer station. They're just curious stories of isolated problems, disconnected from any larger plot. But when they're put together as individual strands in a larger web, a far more worrisome picture emerges.

We know that the above countries and terror groups talk together and trade with each other and have the various parts to pull off a major EMP on an unsuspecting nation. We know some of them have the intent to cause catastrophic harm to enemy nations. The only saving grace is none of them, either acting individually or collectively, have yet to put all of these parts together and act on any foul intentions. Yet.

# THE CHEYENNE MOUNTAIN SAFE HOUSE

If there's one place in North America you want to be close to when the attack hits, it's Cheyenne Mountain. And the only thing better than being close to it is being inside it. Right in the middle of the mountain.

The mountain states of the American West have always carried a certain allure and mystique to them, going back to the founding of the United States. They were home to warring Native American Indian tribes. They were the sight of Pike's Peak Gold Rush in the 1850s, one of the largest gold rushes in the country's history. And they played host to many tales of cowboys, gunslingers and robbers – some true, some the stuff of fiction.

Colorado's Cheyenne Mountain, about 85 miles south of Denver, is a picturesque location. The highest of its three peaks stands 10,000 feet high. These mountains and the luscious evergreen trees encircling them caught the eye of entrepreneurial prospectors. They set up resorts, health spas and colonies. It became both a stopover and a destination in its own right. One of the more luxurious locales in the region was The Broadmoor Hotel, originally built in 1918. It still stands today, boasting 5-star ratings and celebrity guests. One night in 1987, then future president George W. Bush became extremely drunk at the Broadmoor and suffered a wretched hangover the next day. It was one of the worst of his life and, recovering in the hotel, he pledged to bid farewell to alcohol for the rest of his life. And he did and it changed his life.

But for the most part Cheyenne Mountain is no longer the site of such western lore. Now the storytelling that emanates from it is the stuff of conspiracy theories and science fiction movies. Yes, big rocks can carry gold inside and around them. And, yes, they're good places for cowboys on the lam to run and hide. But they're also safe havens for secret military operations. That's clearly what the U.S. military thought looking at those mountains during the Cold War.

As the government developed its own nuclear program they also worked hard on building an early missile warning system to defend against attacks. They needed to monitor the skies to watch for any missiles launched by the Soviets. Few analysts saw the creation of nuclear weapons as an offensive move. They were for defensive purposes. The whole point of the arms race, the whole justification for quickly building nuclear weapons, was to send a message of deterrence. If only one side has a devastating weapon then they've basically won. The mere threat of being able to launch such a weapon is enough to make you the victor. However if both sides have one, it's a check and balance on the other. Mutual deterrence only works though if you can pay attention to what the other side is doing. Thus, watching the skies for incoming missiles was just as important as having your own weapons for retaliation.

The United States first realized missile defense was a priority for them after the Imperial Japanese Navy's surprise attack on the U.S. Naval Base at Pearl Harbor. President Franklin Roosevelt described December 7, 1941, the day of the attack, as a "date which will live in infamy." The attack, in which over 300 Japanese planes launched from six aircraft carriers, killed 2,403 Americans and wounded another thousand. It was the event that pushed the U.S. into joining the Second World War. They'd never been surprised like that before. American military commanders vowed never to be surprised like that again.

"In view of the possibility of air attack in any future war, we feel that the air defense of the United States cannot be left to chance," General Carl Spaatz, commanding general of the Army Air Forces, testified to Congress. "There must be a commander responsible for it. We must be properly organized so there cannot possibly be an air surprise such as occurred at Pearl Harbor." There was already an entity called the Air Defense Command, established in 1939, but it was mostly tasked with investigating the issue. It wasn't doing the actual work. "We hope and expect we will have enough appropriation to provide equipment and personnel to maintain radar stations open 24 hours a day," Spaatz continued in his call to action.

Congress finally heeded his call. After much discussion, the Joint Chiefs of Staff approved the Continental Air Defense Command (CONAD), established in 1954. But soon after it was launched, it became clear Alaska – then not a state – and Canada needed to be included so all vantage points could be properly safeguarded. CONAD expanded to become the North American Aerospace Defense Command (NORAD) in 1958. It served as the keen eyes for Canada and the United States, watching the skies for incoming missiles. It still serves the same function to this day.

By this time, in 1958, it was located in Ent Air Force Base. This was right in the middle of the town of Colorado Springs, which is about an hour's drive south of Denver. There were only 45,000 people living and working there in 1950. But by 1960 it shot up to 70,000. The town was growing, like so many American towns were during the post-war baby boom. Municipal and state services were built. For these reasons and more, it was a good location to host a growing government facility and a place for its employees to raise its families. There were other reasons that made it less than ideal.

The big liability was that it was above ground. Exposed. Vulnerable. If war broke out, it could easily be attacked. A handful of well-trained operatives, like those unidentified figures who attacked the Metcalf transformer station, could storm NORAD and leave North America blind to a nuclear attack, unable to respond. It was an unacceptable risk. A key operation like this needed to stay open no matter how bad the fight. There's no telling at what point during a conflict the enemy could launch a missile strike.

General Earle E. Partridge, the commander in charge, was well aware of this problem and sought a solution. "It has been recognized for several years that the facilities at Ent Air Base are quite inadequate both from a point of view of availability of floor space as well as security," he wrote in a letter to military leaders in both the United States and Canada. "The combat operations center is a concrete block building of extremely light construction and is exposed to the traffic on the adjacent street so that a man with a bazooka passing in a car could put the establishment out of commission."

Partridge's lobbying efforts were a success. But they needed to find the right location. And it had to fulfill a number of criteria the general outlined: it needed to withstand the pressure from a sonic boom or a major explosion; there had to be low seismic activity; it couldn't be too close to any other potential targets; but it also needed to have multiple routes in and out. The U.S. Army Corp of Engineers investigated multiple options, conducting drilling operations and rock and soil tests, before settling on Cheyenne Mountain.

In 1959, the military started on their plans to build the new compound on the mountain located on the south-western edges of town. The skirt of the mountain and surrounding area is known for its natural beauty. Hikers explore the various falls that are within the two state parks at the foot of the mountain. Birdwatchers scout for breeding Virginia

Warblers. Bobcats and mountain lions roam the park. None of this is what excited the strategic planners at the Air Force though. No, what drew them were the massive layers of granite that make up the mountains. It's not that they built their compound in the middle of the Cheyenne Mountain area. They built it right inside the middle of the 100 million-year-old mountain.

Just the planning stage took two years. It wasn't until May, 1961, that the groundbreaking took place, at which generals detonated the first dynamite charges. They were far from the last. The excavation period alone took over a year after that. The miners working on the project used over one million pounds of explosives to take out almost 700,000 tons of granite. That was the easy part. For the next few years, the engineering corps built what is basically a small town within the rocks. The inner complex is made up of fifteen buildings, each of them three storeys. It takes up a half acre. Each building sits on giant springs – a total of 1,000 combined – that are connected by flexible pipes. The buildings can absorb shock and won't move by more than an inch. They're nowhere near the peak of the mountain either, but 2,000 feet below.

At its busiest the community housed over a thousand people, many of whom spent their days watching over the 7,000 aircraft in the sky at any given time. While most of them were Americans, a tenth of the staff were Canadian officers. The workers slept in cots and had access to a store, cafeterias and fitness centres. There's even a medical centre on site. They have everything both for convenience and necessity in the event the complex is locked off from the world for weeks at a time.

After construction was finished and various units slowly moved in, Cheyenne Mountain became fully operational as the NORAD Combat Operations Centre in February, 1967. For almost four decades, the complex watched the skies and operated as frontline surveillance in the Cold War. It captured the public's imagination and was even the setting for the 1983 movie War Games and the 1994 science-fiction feature Stargate.

It was never attacked during those decades, either by physical sabotage, cyber attacks or other methods. At least nothing's been made public. Each of the 15 buildings have 40-inch steel doors meant to withstand the most massive of blasts. They're surprisingly easy to close – a single person can do it by hand – but are rarely shut. The last time they were closed was on 9/11, when they feared the mountain was among the terrorists' targets. Little did they know at the time but that closing of the doors symbolized the looming closing of the entire facility. After 9/11, American military and intelligence forces adjusted their priorities. The Cold War was over. The war on terror had begun.

"A missile attack from China or Russia is very unlikely," Navy Admiral Timothy J. Keating, then the commander of NORAD, told The Denver Post in 2006. They just didn't need to bunker down and watch the skies with such intensity as before. They had a different focus. Besides, maintaining and coordinating multiple facilities in the region was logistically challenging and costly. While the actual people and devices monitoring the skies had moved from the core of Colorado Springs into Cheyenne Mountain in the 1960s, many additional services had remained in town, at the Peterson Air Force Base. The operation was split across town and they figured it was time to consolidate as much as possible.

As the Washington Post explained in a July 29, 2006 story: "The commander of NORAD works from Peterson Air Force Base, and the trip to Cheyenne Mountain can be time-consuming if traffic is bad. On Sept. 11, 2001, Colorado newspapers have reported, the commander spent 45 minutes on the road between his office at Peterson and his communications center under the mountain while the attacks on the World Trade Center and the Pentagon were taking place."

It was at this point that operations came out of the

mountain. Military leaders in charge did make it clear the Cheyenne facility would remain on "warm standby". This meant they could fire it all up again with only a few minutes notice. It was a wise move. Because it didn't take long before they decided it was time to return to the mountain.

In 2015 the Pentagon announced they were hauling their tracking and communications equipment back into the Cheyenne complex. On April 7 that year Admiral Bill Gortney, the commander of NORAD, gave one of his first big press conferences on the job. He'd only been at the post four months and took to the Pentagon Briefing Room to answer questions. The combination of topics discussed was seemingly broad. Yet clearly connected. They provided a window into the capabilities of NORAD to address international threats and just what those threats were at the time.

Can the U.S. stop a missile from North Korea? one reporter asked. Gortney was confident they could, but he also wanted to extend their reach even further. "We need to be able to start knocking them down in the boost phase and then – just after that, and not just rely on the mid-course phase where we are today," he said. They want their warning systems to kick in even before the missiles are launched. Why not get notification once the launcher is powering up? The more lead time the better.

"In the case of Iran," Gortney continued, "which we don't think they have the capability of today, but what if we got that intel wrong and they moved that delivery capability to the left, even if they moved it today, we could defend the nation with what we have today." Basically, here Gortney says Iran doesn't yet seem to be a threat but if that suddenly changes they're confident they could shoot the missile down.

There's a curious piece of trivia to go along with Gortney's briefing. It happened to coincide with a power outage in the Washington, DC, area. The streets of Washington and

Maryland briefly descended into chaos. Government workers poured out of their suddenly darkened offices. Traffic lights and subway stations shut down. The Secret Service, unaware of the cause, shifted into a high state of alert.

The Briefing Room wasn't impacted. Everything was fine where the press conference took place. Yet this didn't stop reporters from asking Gortney his perspective on the issue. And the admiral, like everyone else, didn't know the cause. But he did stress the importance of the grid:

"What it really goes to is we have a lot of vulnerabilities out there... It is our reliance on critical infrastructure that our nations need in order to operate, be it banking, be it power, be it rail, be it the FAA, and if someone, either through a nefarious act or just through an act of nature – they impact on us. And so I think all of our - those critical infrastructures are - are fragile. And when I say fragile, it's just because we really don't know the true vulnerabilities. We try and mitigate them as best we can. But it causes me great concern," Gortney said, according to the unedited transcript. "If the power grid up in Ottawa fails, then we - that could take the northeast quadrant of the United States out. Our interdependencies, not only within our own country but the close linkages between us and Canada, those are all - and it's not just limited to power. It's also limited to everything else that we rely on for our governments to run and our countries to run."

This tells us that Gortney, the top military authority tasked with watching the skies for an attack, was well-versed in the interdependency and vulnerability of the electricity grid. He was clearly concerned about it. Even though conventional wisdom tells us NORAD's main task is protecting the skies from bombs hitting North America. That's their job. The electricity system shouldn't be any of their business.

A few minutes later though, Gortney revealed likely why it was that grid protection – on that day of a major power outage in one of the most important regions of the United States – was so front of mind for him.

A keen reporter brought up the fact that in the March 30, 2015 public notice of contracts awarded by the U.S. Department of Defense, a company won the tender for a $700 million contract for work at Cheyenne Mountain. That's a lot of coin for an outfit supposedly on "warm standby." Contract release CR-058-15 revealed that Raytheon Technical Services "will provide sustainment services and products supporting the Integrated Tactical Warning / Attack Assessment and Space Support Contract covered systems." The reporter asked what this all meant.

Gortney was not a politician. So he didn't dodge or dither. He just flat-out answered the question. Here's their exchange in full, taken from a Department of Defense transcript:

Q: Can I ask you about the Cheyenne Mountain complex? Several years back the NORAD and NORTHCOM moved a lot of its command into – at the Peterson Air Force Base. However, in the last week, there was a $700 million contract let for activities at the mountain. Has there been any change in the status, and do you know what that $700 million's going towards?

ADM. GORTNEY: Yeah, we're – there's – because of the very nature of the way that Cheyenne Mountain's built, it's EMP-hardened. And wasn't really designed to be that way, but the way it was constructed makes it that way. And so, there's a lot of movement to put capability into – into Cheyenne Mountain and to be able to communicate in there, and that's what that contract's letted for in order to do that. We have the space for it, we have the cube. My – my primary concern was do – are we going to have the space inside the mountain for everybody who wants to move in there, and I'm not at liberty to discuss who's moving in there, but we do have that capability to be there.

Q: (off mic)

ADM. GORTNEY: Well, it goes to the very nature of an EMP threat, I think, that – that capability that we need to be electro-magnetic pulse be able to sustain those sorts of capabilities, our ability to communicate, things of that nature, and an EMP environment's important. [Author's note: It seems here that Gortney, speaking jilted and off-the-cuff, is saying that being aware of the EMP threat and protected against it is important.]

Q: Are operations still going to go on at Peterson?

ADM. GORTNEY: Oh, absolutely. We'll be in both places. We – we – we command where the staff is, and we move between both locations so that we can co-opt, should we need to, both NORAD and NORTHCOM. We're going to maintain both.

Q: Just a follow up on that question. How soon are you looking to move some of those capabilities into that complex or —

ADM. GORTNEY: Well, we're – it happened long before I got there, the people are moving in there. And so it was, you know, decisions from my predecessor and I support those decisions. And we'll make sure that it all gets in there and it's all secure.

This changed everything. It was a rare moment for the handful of people who'd been trying to make EMP a mainstream conversation. They'd faced a lot of challenges over the years. Good luck getting officials to answer questions about it. Good luck wringing anything from government media liaisons, if they even understand the question. And don't even bother putting any questions to politicians, most of whom are clueless on the subject.

However on this spring day, a new NORAD commander standing up to conduct his first open topic press conference on the day of a Washington black-out simply volunteered startling information on EMP. He wasn't even asked. Just like that, he spoke about it openly. Gortney hadn't been on record saying anything on the topic before hand and didn't publicly say much more of note after the fact. He then retired in May 2016.

What he did say that day though was enough to make it a watershed moment: Not only was the military aware of and concerned about EMP attacks, but they were spending hundreds of millions of dollars to relocate missile defense equipment to protect it from just such an attack.

Gortney's comments weren't the first time though that the government acknowledged there was something especially attractive about the Cheyenne Mountain complex aside from the physical security provided by all that granite. Few members of the public have visited the facility. It's not like it has open hours like the White House. On the NORAD website's page about the complex, they discreetly but clearly note at the bottom that "public tours of Cheyenne Mountain are not available."

In the summer of 2009, technology journalist Daniel Terdiman was granted a rare visit to the complex. He was even allowed to take a camera inside and photograph parts of the facility. The unspoken assumption in Terdiman's feature article for CNet is that he was allowed in because it wasn't fully operational. They'll be less forthcoming now that they're watching the skies once again. And will they be so eager to hand-out the fact sheet Terdiman received then?

"According to a fact sheet I was given, the threats that the MSG [Mission Support Group, which managed the complex while on "warm standby"] is geared up for, in descending order of likelihood, but increasing level of consequences, are: medical emergencies, natural disasters, civil disorder, a conventional attack, an electromagnetic pulse attack, a cyber or information attack, chemical or biological or radiological attack, an improvised nuclear attack, a limited nuclear attack, or a general nuclear attack."

It's a shocking piece of reporting that remains hidden in plain sight on the popular online technology website. Shocking and highly informative because it reveals several pieces of previously unclear information. It says NORAD is aware of EMP. It says they're concerned about it enough to place it on their list of credible threats. It also, perhaps most astonishingly, places it higher on the list of likely attacks than cyber, biological or general nuclear warfare.

Rather oddly, at least according to Terdiman's representation of their list, they also place it as of less consequence than these attacks. A targeted EMP attack covering a limited geographic region - such as one designed to only impact Cheyenne Mountain - is certainly less of an issue than dropping a nuclear bomb on the ground. But a much broader EMP pulse originating significantly higher in the atmosphere is, at least in the long run, much more dangerous than a traditional nuclear bomb. Yet threat assessments can be just as much an art as they are a science. The key fact here is that EMP even commands this placement on such a list in the first place.

What does Gortney mean when he says the mountain is EMP hardened? And that this is not so much because of how it was designed but constructed? Terdiman's article, written six years before Gortney's press briefing, also touches upon this key question: "the complex is set up to shield the interior against an electromagnetic pulse (EMP), which can fry most electronics. [The civil engineering director for the site] said that, in fact, Cheyenne Mountain is the only DOD high-altitude Electromagnetic Pulse certified underground facility. Among the provisions are wall-mounted EMP filters called metallic-oxide varistors, which dampen the pulse, as well as a system that allows personnel inside to break away interior electronic systems from the external commercial power systems."

These are also startling revelations, given the source. It's telling that the head of engineering let a journalist leave with the impression that, yes, an EMP can indeed fry most electronics. It's not that this isn't true. The evidence is foggy due to the limited testing, but it certainly leans towards extensive damage. It's just interesting that senior defence experts clearly aren't downplaying the potential severity of an attack. They clearly believe the worst case scenario, however improbable, is still possible.

The statement tells us other things too. Like that the Department of Defense classifies facilities on whether or not they're EMP certified. It also shows that they're operating with the belief that even though their external walls may be shielded against the pulse, that doesn't mean the pulse won't still come in via the electrical lines. Safeguarding against this is one of the benefits of having the ability to break their internal systems away from the commercial grid.

It's alarming to discover that the U.S. military and NORAD do take the more severe consequences of EMP as a very real possibility. In some respects, it would be nicer to think that this whole story was indeed nothing more than the musings of conspiracy theorists or old-fashioned ideas from the Cold War that have since been abandoned or debunked. The truth is just the opposite. Senior military experts have clearly assessed the risks of danger and consider them significant enough to justify a major relocation of their equipment and personnel at significant financial expense.

What's no less telling is the timeline. The collection of knowledge about EMP hit a peak in the 1960s, before test ban treaties banned useful field research. For years these ideas essentially collected dust. Even in 2006, when NORAD functions were taken out of the mountain, the issue clearly wasn't front of mind. Or else they wouldn't have exposed their operations to this risk in the first place. But something happened in the decade that followed, something that made military experts change their minds.

What happened that made NORAD decide EMP was a

serious concern? Maybe they just knew so little about it in past decades. After all, as we will discuss in future chapters, the issue was largely classified until 2008. But maybe it also had something to do with North Korea's nuclear tests and the launch of the satellites, like the KSM-3. Maybe it was connected to Iran's increasingly obvious nuclear intentions. Or maybe it was all the weaponry being traded between countries like Russia and rogue states and terror cells.

We know now the Cheyenne Mountain complex is the ultimate safe house. But that in itself is not enough. The rest of North America is not EMP-hardened. Regular industrial, commercial and residential facilities aren't at all constructed with a view to protection. It's not included in state and provincial building codes. What good is a functioning base if the hundreds of millions of people it aims to protect are plunged into darkness and anarchy? The good news is there can be more safe houses. Many more. In fact, all of North America can be turned into a giant fortress to fend off EMP attacks. Resilience matters. And it's possible.

But before we get to the story of how to protect the continent, we first need to meet the people working tirelessly to do it.

# THE GREATEST STORY NEVER TOLD

Try to come up with a picture in your head of the type of person who would sound the alarm about the dangers of electromagnetic pulse. What would the stereotypical EMP warrior look like? No doubt what entered many people's heads was an older, grizzled man, stern and no-nonsense, a veteran of the army or the FBI or the CIA, a died-in-the-wool Republican to boot.

He'd be someone who delivers secrets in brown envelopes transferred discreetly in parking garages. He's a tough guy who ferrets out the truth against the odds. Someone who's been around the block enough to be jaded about how the system works but still hasn't thrown in the towel. Like an iconic figure from detective fiction or conspiracy movies. Good guess. But entirely wrong.

Andrea Boland is none of these things. And yet she's the gold medal champion of the movement, the first politician in North America to bring about legislation aimed at protecting her nation from a catastrophic EMP attack.

Photos on Boland's website show her posing with her dogs, attending local farmers' markets, holding up hand-made quilts and enjoying the picturesque scenery in the southwestern region of the state of Maine in which she lives.

Before getting into office Boland worked as a paralegal and most of her political activities were community ones, like her involvement in local school issues and writing letters to the editor in the local newspapers. It was happenstance that she even stumbled upon the issue of EMP in the first place.

Boland first ran as a Democrat in the Maine House of Representatives in 2006, winning a two-year term. It wasn't until 2011, while she was speaking with her science advisor about cellular radiation, that Boland first even heard of EMP and learned about the havoc it could cause. It clicked with her right away. "I was just aghast that we could allow this extensive grid we had that we were so dependent on to be so vulnerable," Boland told me one day in late 2016. "I just thought if there's anything I could do, I'd try to do it."

It didn't take long for Boland to turn concern into action. "I've always felt that if I'm going to be in the legislature I might as well do something helpful," she said. Within months of first studying the issue, Boland put together legislation designed to protect the Maine state electric grid from EMP. The first attempt, in the fall of 2011, all happened so quickly that legislators didn't have time to grasp the complexities of the issue and the bill failed to pass. Yet this wasn't the end. Far from it. Andrea Boland's fight as an EMP warrior was only just getting started.

Her quick, nimble and unapologetic legislative push impressed the small community of international experts who actually knew about EMP and considered it a priority. They invited Boland to join them in London, England, for the Electric Infrastructure Security Council's annual summit. It's the "everybody's who's anyone" event of the EMP community, which sounds like it would be a big show but is in fact a small club. What it does have going for it though is the impressive credentials of its attendees.

Right Honourable. General. Chairman. Lord. Doctor.

Professor. These titles and others were affixed to many of the name tags of the less than one hundred people who gathered in the august halls of Westminster Palace in the spring of 2012 to discuss their plan to save society from imminent doom.

"If we're successful here, probably no will ever even be aware of our efforts. We won't get any praise for it," the chairman of the summit said that day, Boland recalls. "But if we aren't successful, this will be just the end of life as we know it." They were the front line warriors in the greatest story never told.

They were all told to look around the room. And what would they see? The only people in the world who were truly working on the issue. A topic so important, with stakes so high, and only eighty or so individuals cared. For all the thousands of research papers and books and conferences and discussions and committees convened for the general topic of nuclear weapons or for, say, world hunger or climate change, this was it for EMP.

A meeting like this should have garnered major attention. Everyone should have been talking about it. Dozens of prominent figures, meeting in a major venue to discuss a once classified international security issue. It almost sounds like the meeting of a secret society, of shady figures controlling the future of the world from the comforts of their boardroom. Did the press beat down the doors to shine a light on this meeting? Did they stalk the attendees in the hallways, eager to reveal what was happening behind those walls? Hardly. The media attention was almost non-existent.

The sad fact is that this was no secret society. It wasn't hush-hush at all. There was nothing the decorated attendees would have loved more than to speak to the world via television cameras. This was not a battle they planned to fight via back channels. They weren't trying to keep the EMP community secret. They wanted it out in the open.

"We would have loved coverage," Boland recalls. "There were amazing things being said by serious people. And I knew them to be true by the work I had done. And the media didn't want to cover it." There were a couple minor mentions on the first day of the summit and then that was it. No follow ups.

This didn't deter Boland. The summit inspired her further and she hit the ground running at the next session of the Maine House of Representatives. In February 2013 Boland presented LD-131, a piece of legislation that, unlike her last effort, got to the floor of the legislature. It was the first ever EMP bill to make it to a vote at a state legislature anywhere in the United States.

Boland knew by then that anyone uninitiated in the EMP conversation – which was basically everyone in the legislature and everyone who followed state issues – approached the unfamiliar subject with caution. She didn't want to scare anyone off with an overly ambitious request. For this reason, the bill didn't seek to tinker with every single piece of technical gadgetry that fell under the domain of the state. Instead, all Boland proposed was that electric utilities include EMP protection devices in their design and construction. And even then, only new ones. Electrical transmission lines, stations and power plants currently in operation were entirely left alone. There'd be no overhaul of the entire grid. Rather, slowly but surely, the grid would become resilient as its various parts were replaced over the years.

As the beginning of the LD-131 summary reads: "This bill requires a person submitting a petition to the Public Utilities Commission [PUC] for the purposes of receiving a certificate of public convenience and necessity for building a transmission line to include a description of design measures to be used that limit electromagnetic field levels and ensure the protection of the transmission and distribution system against damage from an electromagnetic pulse or geomagnetic storm." If you want the state government to approve your new power lines or generating station, first prove to them that EMP protection is in the plans. It was that simple.

The bill went on to instruct PUC to develop technical standards to govern exactly how they'll determine effective EMP mitigation for the power industry. If you're going to ask people to explain their efforts, you have to lay out some criteria about what amount of protection is considered a passing grade.

The bill was essentially to get the government and relevant industry to officially acknowledge that EMP existed and that they were going to work towards preventing its effects. At the time, a company called Central Power was undergoing a $1 billion expansion project to add more transmission lines and substations to further serve its 500,000 customers. It was the sort of project LD-131 stood to impact. The bill certainly would have added a regulatory burden for the power company. But as far as government legislation goes, it wasn't as onerous as it could have been, especially given the severity of the issue.

Such a modest approach, where very little is actually being asked of the stakeholders, should all but guarantee victory. But in the world of politics, where various interests collide with multiple stakeholders, nothing is that simple. Boland was about to get a crash course in how lobbyists work their trade.

First, LD-131 went to a committee. A number of players, including PUC got involved, and made minor amendments to the bill. Electric companies met with committee members to answer questions about their awareness and preparedness for the topic. The answers, if they even had any, were alarmingly slim. Due to this, the committee revised the text of Boland's original bill. They made it more ambitious and included the requirement that PUC study the whole state's vulnerability to EMP. They then sent the bill to a vote. It passed the Maine House of Representatives. And not by a slim margin either. By a unanimous vote.

The bill then went to the Senate, where it again passed without a single opposing vote. Then it came time to cross the desk of Maine Governor Paul LePage. This was where Boland and her allies were nervous. LePage hadn't been subject to the same personal lobbying and information campaigns as Boland's colleagues. It was unclear what he'd do. The governor's three options were to sign it, veto it or let the time window elapse whereby, if he does nothing, it automatically becomes law. LePage picked the latter. In June of 2013, Andrea Boland successfully shepherded into law the very first legislative measure to protect against EMP in North America.

Did that mean victory? Not so fast. The big hurdle was the report by the utilities commission. While you'd think they have the interests of ratepayers and citizens front of mind, they're also players in a major industry. The North American Reliability Corporation (NERC) is a not-for-profit regulatory authority tasked with assuring the reliability of the grid. While it would make sense for them to be natural allies with the likes of Boland in the EMP battle, they've actually been critical of protective measures. Their past positions have focused on their belief that EMP protection is an issue of national defence and therefore the responsibility of the military and federal government.

To add to the confusion, NERC acts as both the de facto lobbyist for power companies as well as the group that sets industry standards. That's quite a conflict of interest. William Graham, Ronald Reagan's former science advisor, doesn't mince words when it comes to the regulatory group, writing in a foreword to Peter Pry's book *Blackout Wars*: "The Federal government is failing spectacularly to protect the electric grid and the American people from nuclear EMP, or from any of the other catastrophic threats, because of a self-interested and wholly improper relationship between the electric power industry represented by NERC, and the U.S. Federal Energy Regulatory Commission (FERC)."

No company wants to face added restrictions, but NERC's relationship with the government commission that overseas energy is seen to make this dodging of regulations easier. Even if they acknowledge that the concerns are valid, they'd much rather promise to take care of things on their own than follow the rules and timeline of government legislation. So NERC is viewed as an authority by power companies across North America and viewed with suspicion by the experts eager to introduce EMP legislation.

By the time the report mandated by LD-131 came in from the utilities commission, the fingerprints of NERC were all over it. Instead of detailing the extent of the EMP risk and how it could be mitigated, it explained the problem really didn't lie with the state government. It's a line of argument they'd used before. This ground everything to a halt. It was assumed by Boland and her allies that the report would include recommendations to enact industry standards for protection. Instead, the report passed the buck. So while the bill itself was passed, the report it mandated failed to offer a plan of action, effectively throwing the issue into limbo.

In the meantime, Boland's term ended and she couldn't seek re-election because of term limits. But EMP protection was no make-work project for her; it was an honest cause. She didn't drop her advocacy. Boland partnered with a state senator in 2015 to propose another piece of legislation that directly instructed two major power suppliers to protect against EMP. The bill was defeated.

Looking back on it all, over a year later, Boland is serene and optimistic. "You seem to grasp at every little bit of progress but I guess that's how the world often works, just in small steps," she told me. "I don't expect big things to happen fast because there's so much threatened by this vulnerability, but I'm hoping that we have progress on it."

"The information is out there if people go seek it out and if they can believe that some of these people that we take as authorities, like utility companies, aren't necessarily all that authoritative on everything and there may be something there that you really have to explore by yourself," says Boland.

However, her efforts weren't for naught. The power companies in Maine clearly weren't pleased with the idea of EMP-related regulations. But shortly after the defeat of these bills, they took it upon themselves to begin minor investigations and parts replacement that started them on their way to EMP protection. Maine is better protected now than it was several years ago, all because of Boland's efforts.

In the 2016 American elections, Boland unsuccessfully ran again for a state Senate seat. Despite the legislative and electoral setbacks, she plans to continue her advocacy to protect her country from an EMP disaster.

If Andrea Boland is the opposite of what comes to mind when you first think of an EMP warrior, Peter Pry is right on the money. He's a former CIA agent. Conservative-minded. He's quite tall, wears black suits, and has a firm handshake. Pry would fit right in as a character in a Washington espionage thriller. He's also got quite the nickname used by those who follow the issue: Mr. EMP.

Pry's the man who got Boland hooked on the subject. He first introduced James Woolsey, the head of the CIA during the early years of the Bill Clinton administration, to the topic. He also taught dozens of members of the United States Congress about the issue. There is nobody in the English-speaking world who has done more to educate people about EMP than Dr. Peter Vincent Pry.

"My involvement in this topic starts from before I was born, in the Russian Civil War," Pry told me one day, in a sit down interview, in the fall of 2016. Pry's great-grandfather was a general in the Czarist army that fought alongside the anti-communist White Guard against the Bolsheviks. The anti-communists lost and pieces began moving into place for the eventual creation of the Union of Soviet Socialist Republics. But before that could happen, many anti-communists smelled the winds of change and fled. Pry's family left for the United States in 1919.

"Anti-communism was like a religion in my family. We'd lost people. Family fortunes were destroyed," he explained. "My parents were very concerned that someday the Reds were going to come here." So his uncle and father fought in World War II. "We were raised with that responsibility," said Pry. "It was up to us. Freedom could be lost in one generation."

While growing up in the 1960s, Pry became fascinated by military history. He read up on ancient and current eras. He studied military doctrines. Then, at 13-years-old, he was transfixed by a book by the American military strategist Herman Kahn called On Thermonuclear War. While it was a technical title from an author who was, at the time, a think-tank staffer, the book sold an astonishing 30,000 copies shortly after it was published in 1960. It drew attention for coming up with the concept of the Doomsday Machine, a hypothetical nuclear weapon that could destroy all life on earth.

"What I learned then was that the most decisive military technology of the age is what usually determines the future," recalls Pry. "I decided then that that's what I was going to do, go into the military or become a civilian expert."

But before Pry could become Mr. EMP, he first had to learn about it. The news came by way of a textbook Pry happened upon at a college library in Utica, New York, the town he grew up in. Samuel Glasstone was the author of dozens of popular textbooks on nuclear, chemistry and space topics. But none were as influential as The Effects of Nuclear Weapons, which first came out in 1950. It included a chapter on EMP.

The young Pry was so transfixed by the book, he actually

stole it. He now admits to the petty theft with a chuckle. Years later, he was speaking with a college professor in the same community who by chance related a conversation he'd recently had with a local librarian. She had grown concerned that the government was running a program where agents were traveling around the country confiscating books that revealed nuclear weapons and national security secrets, because one day a book that nobody had ever once loaned from the library – Glasstone's – had mysteriously vanished.

Once Pry got to college, he was even more committed to military strategy, going so far as to complete two PhDs on related topics. The subject of EMP wasn't often front of mind for him during those early years. He knew of it, as many do, but thought of it as something of a side issue. But it kept cropping up.

One of the visitors who stopped by his class was William Graham, who would go on to serve as President Ronald Reagan's science advisor. Years before though, Graham was one of the people sent to Hawaii to investigate the relationship between the Hawaiian streetlight incident and Starfish Prime. Graham was one of the original front line warriors in EMP.

After academia, Pry entered the CIA in 1985 and it was there that he'd have an experience similar to Graham's. One of his first assignments was to figure out why the Russians, after signing various arms control treaties, were so eager to hang on to a bunch of missiles called the SS-18 Satan.

"I was looking at this and wondering: what can you do with these missiles? There weren't many things they could do with them. The only thing that made sense was EMP." The SS-18s are intercontinental ballistic missiles with a massive range. One variation of it can travel 16,000 kilometres. The distance from, say, Moscow to Washington is 7,800 kilometres. An EMP missile, they then realized, was basically a slow-burn Doomsday Machine. And the 'Evil Empire', as Reagan called the USSR, had Intercontinental

Ballistic Missiles that they could point it on.

"Most of the people I was working with at the CIA had never heard of this before," Pry said. "They were astonished when I brought it to them. It was a deeply classified, very technically oriented sub-set of nuclear effects." The big problem with being a researcher or agent in the CIA, is that your job is simply to observe and report. Recommendations are left up to other people. Your influence is limited. You can write the most groundbreaking report ever, but if the person one level above you lets it collect dust on his desk, then everything grinds to a halt.

Following a decade at the agency, Pry switched to a job where he could make recommendations. He became a senior staffer at the House Armed Services Committee, working with the dozens of members of Congress who oversee the nation's military and its research and development. There he took on many tasks, including researching prospective North Atlantic Treaty Organization (NATO) member countries' militaries. If there was a group of people to educate about a potential catastrophic attack on the homeland, a group that needed to hear about the importance of EMP, this was it. Pry was in the right place.

Yet it wasn't his efforts that eventually swayed them to understand the problem was real. It was the Russians. A U.S. congressional delegation went to the Balkans during the war in Bosnia, which was the first time NATO engaged in combat operations in its history. The Russians had always responded to NATO with caution, viewing it at times as downright hostile. After all, it was created to serve as a check on Cold War military aggression.

"You know, we're not helpless to stop you," was along the lines of what the Russian delegation said to members of Congress, according to Pry's retelling of it. They were talking about NATO's presence and potential expansion. "All we have to do is launch one ballistic missile over your country and conduct an EMP attack and that would be the end of America."

That's right. Russian officials openly teased a group of American politicians about destroying their civilization right in front of their faces. As soon as the American delegation returned home, word got around. Pry had been seeking approval to set up a commission on EMP and, after that incident, the Democrats and Republicans voted for it unanimously.

It was then that Pry shifted gears from the various different topics he focused on as a staff member of the House Armed Services Committee to focus exclusively on the EMP Commission. This was a multi-year project that looked at the subject in greater detail than had ever been done before. Its specific task was to grapple with how an attack would impact the United States, what could be done to prevent and recover from an attack and what action was needed to get these results.

The Commission's primary accomplishment was released in 2008 – the 200 page Critical National Infrastructures Report that now serves as the bible of official EMP data, the only U.S. government document of its kind. The chapters cover extensive territory on the impacts an attack – natural, like the Carrington Event, or man-made from an enemy – will have on various sectors like banking, natural gas, aviation, food infrastructure and so on. It identifies the problems and proposes solutions. It was a major undertaking. Yet, much to everyone's frustration, it accomplished little. "It ended, we made our recommendations and unlike previous commissions, they weren't implemented," says Pry. "We didn't want it to end until the problem was solved."

This lack of concrete action wasn't for lack of trying. Shortly after the report, there was a brief surge of political momentum. The Grid Act and Shield Act were two bills that had bipartisan support that were stalled or killed off through various legislative methods. Pry believes with certainty, like Boland, that the deep pockets of the utilities lobby played a part in burying these bills. The EMP Commission was shut down for a number of years. And those involved kept up their work on a voluntary basis, that's how much they were committed to the cause. Now additional funding has been secured and the commission is back in action.

In the years after the report came out, Pry developed his reputation as Mr. EMP, traveling to conferences, meetings with legislators and drumming up public interest. It sounds dramatic and exciting. And in many ways it is. But taking the path less travelled has not been an easy journey for Pry and his fellow EMP warriors.

It was on a fall day that Pry and I had one of our meetings, this one at Massey College in Toronto. He was there to speak at a cyber-security conference happening at the respected and influential college. As we sat inside talking we could see students coming and going, locking up their bicycles as they left for class. These students will go on to hold powerful roles in society, as leading academics or lawyers, business executives or perhaps even a prime minister or two. But just like Boland's conference in Westminster Palace, the students weren't aware of the important event happening right in front of them. No one was, aside from the attendees. People weren't clamouring to get in. Nobody from the media – except this author – was there, beating down the door to learn about these urgent issues.

"It's very politically incorrect to be trying to raise awareness about EMP," Pry said, reflecting on his life's work. "All of us have made sacrifices." If Pry had followed a typical career path and championed the issues the administration of the day wanted him and his colleagues to follow, he'd be enjoying his 60s with a very well-paid position at a large defence firm by now.

But he became more and more vocal during the Obama years about the urgency of the topic. He put his name to op-ed pieces appearing in major newspapers like The Wall

Street Journal. He appeared on television news programs. Ultimately, what he was doing was pointing out a weakness in government policy priorities. The government didn't like this. They started cancelling the contracts he had with them.

"I ended up losing half my income, one by one, as I wrote articles critical of the administration. I was told by the contracting officers, that this was happening because of my insistence on making it a problem."

Pry even lost the family farm. Literally. He and his wife had bought an abandoned tobacco farm and turned it into a 100 acre horse farm. But they lost it in 2014. "We struggled to try to hang on to it," he recalls. "It still hurts to think about it. It hasn't been a rewarding life. This is the one thing I could have given her and we had to lose that."

There have been successes, certainly. The Obama administration, Pry says, agreed to give the money to harden Cheyenne Mountain. It was a minor victory, but one that at least shows an administration subtly acknowledging both the problem and the need for solutions. Plus there are all the attempted pieces of legislation at the federal and state level. None finished the job. But progress has been made. It's been quite a slog though for the handful of hardened EMP warriors out there.

"Many is the time I wished I'd never heard of EMP so that I wouldn't have to do this. But I do know about it and I don't have any choice now. It's been a pretty thankless life." It was in 1995, when he transferred from the CIA to Congress, when Pry planned to start really doing something about the issue. He thought it would take a year or so to get legislation in place to protect the nation and fulfill his life's work. Then he could move on to other projects, other passions. But hurdles arose one by one and the years went by. "Twenty one years. My god. I've become an old man and we're still not protected."

108

# PROTECTING THE GRID

Despite the best efforts of people like Andrea Boland and Peter Pry, North America is still vulnerable. The electrical grid is not resilient. It could go down at anytime.

Slowly though, more and more people are waking up to the issue. There will come a time, hopefully sooner than later, when public attention hits critical mass. What happens then? What real world things need to happen to protect the continent from catastrophe? Can it even be done?

Yes it can and it's surprisingly simple. For all of the talk of classified files and clandestine tests and international intrigue, the answer to this seemingly insurmountable problem is almost something of a letdown.

The stakes are so high and the fallout from an attack is so dramatic, you'd think there would need to be some herculean effort to prevent it. Billions, perhaps even trillions, of dollars spent. Massive overhauls of physical infrastructure. A complete rewiring of each and every one of the products that makes up our electronic civilization. Not at all. The United States and Canada really just need to replace and add parts to their utilities systems to the cost of a few hundred million dollars. That's it.

The EMP Commission's research concluded that protecting the grid for the United States would cost the government $2 billion. For perspective, the U.S. federal budget for 2016 was $3.5 trillion. This means the costs to federal taxpayers to protect their entire civilization from downfall is just 0.0005% of one of their annual budgets. And it's not like all the expenses need to be rolled in a single calendar year.

These costs aren't entirely borne by the feds either. Many are state costs. The average cost per state, when broken down that way, would be a mere $40 million. Governments routinely throw this sort of money at projects of lesser urgency.

What will EMP expenses actually go towards? If the state legislation in Maine kept picking up steam, it would have resulted in legislation directing energy companies to install specific parts to their high voltage transformers and other pieces of critical infrastructure.

There are a number of online stores offering EMP-related protection devices. Many of these are marketed for individuals worried about the impacts EMP will have on their homes and families. A lot of people in the preppers community are experts in these products. They're the surprisingly large number of thousands of North Americans who believe some sort of catastrophic event is on the horizon and want to be prepared.

A number of these products come in the form of boxes or bags made with material that blocks out an electromagnetic pulse. There are entire books out there for preppers and other concerned families to guide them in making their lives resilient. Arthur Bradley is one such author who's written a number of these titles, including Disaster Preparedness for EMP Attacks and Solar Storms. Bradley's no random prepper either. He has an engineering doctorate and works for NASA.

No matter how prepared the individual makes their family

and home, no matter how resilient they become, there can be no mistake that hardening the grid is the top priority for society as a whole. It's the most effective way to guarantee widespread resilience and avoid societal collapse. In the age of the electronic civilization, it is increasingly difficult to live fully off the grid. And a growing number of firms are waking up to the urgency of this issue. They're designing products, filing patents and marketing their wares to governments and major infrastructure companies. North America's vulnerability to this threat is certainly a liability. But it's also become a major business opportunity.

The Pennsylvania-based company ABB Incorporated created a product they've given a mouthful of a name to, the "SolidGround GIC grid stability and harmonics mitigation system". In this case, GIC stands for Geomagnetic Inducted Currents. The phrase refers to the massive currents that make their way into power lines immediately following a natural or man-made EMP. While the product doesn't shield devices from getting hit directly by EMPs that make their way through the atmosphere, it does protect indoor devices from getting hit by a surge coming in through the grid.

The system, their brochure explains, "automatically blocks DC currents from flowing in the neutrals of large power transformers. The system protects the grid against effects of solar storm Geomagnetic Induced Currents (GIC) and Electromagnetic Pulse (EMP) induced currents... Grid voltage collapse and component damage can occur during modest to severe solar storms. SolidGroundTM installations will prevent this collapse and eliminate damage to power system components and customer equipment."

It's a massive piece of hardware, weighing 8,200 lbs and standing 12 feet tall. And the company is no fly-by-night basement outfit, either. They offer various other products, have branches and sales offices across the world and are publicly traded on the New York Stock Exchange. Grid protection is a serious, growing sector.

The Minneapolis-based company EMPrimus holds dozens of patents around the world for grid resilience products. One of their products is EMP Alert, a small mounted device that looks like a large security camera. The problem, the company explains to their clients in a document, is that "without any warning or detection, an electromagnetic pulse can severely disrupt data centers and control centers causing damage ranging from covert data corruption, to computer outages, and destruction of electronics."

This product, they write, "is a low-cost detection solution that can provide an early warning for data and control center targets. It has the potential to detect a threat in time to take defensive action and, as part of your security system, can help locate and eliminate it." They're clearly catering their sales pitch to companies concerned about problems like the Maroochy water incident: "Electromagnetic weapons have evolved from experimental novelties to sophisticated threats. They have now become portable, affordable items that can be effectively employed outside a building to damage all electronic circuitry in data and control centers, including support equipment, cooling, fire and smoke detectors, security systems and back-up generators. They are readily available to terrorists, disgruntled employees, corporate protesters, or anyone intent on causing major disruptions to your business and your customers."

These aren't the only companies out there in this line of work either. There are others, large and small. Some are consumer-oriented, others are geared towards providing products to large critical infrastructure projects. The solutions are available. The products are out there. It's just a matter of government and industry setting standards and having the applicable companies follow them.

Right now, the scant protection that a select number of utility companies are opting for is entirely optional and patchwork. Their industry organizations – like NERC – follow initiatives like Congress' Grid Act or Boland's LD-131 closely. This is no doubt why some are choosing to protect on their own. They want to be ahead of the curve. One report indicated that most utility companies could pay for grid protection by passing on the entire costs to the ratepayers in their bills. This fact will initially worry families already struggling to pay the bills. The final tally though, according to this report, is 20 cents per customer per year.

It's unclear if the costs are the main sticking point keeping the utilities from sitting at the EMP table as an eager participant. Perhaps they're simply stuck in their old ways and want to avoid the effort associated with protection. They want to ignore the whole problem. Or shrug it off. And even those that have sought protection may be taking the cheap route out or simply using products that aren't actually going to do the job. It is for these reasons, and because of the seriousness of the issue, that the likes of Peter Pry argue for national standards and regulation.

Before any of this happens, there are quick fixes out there the EMP Commission recommends. Any U.S. state or Canadian province or company on either side of the border can take the initiative and start protecting against EMP immediately.

The one device that is the best protection against an electromagnetic pulse was created years before anyone had ever heard of the phenomenon. It was even created before the Carrington Event. In 1836, English scientist Michael Faraday was investigating electric fields. He'd discovered through his tests that electric charges only show up on the exterior of a metal conductor. They don't impact the interior. If a box of metal is electrified, the excess will spark off and appear on the outside, causing potential harm. But there won't be any effects felt on the inside. To prove this, he created what's now known as a Faraday cage.

Large-scale versions of these cages are a key quick fix in EMP protection. Creating such a cage is no large feat of engineering. Everything needed to build one can be found at a neighbourhood hardware store. The most rudimentary versions involve placing metal insulation or wire mesh around the subject.

Rather amusingly, Faraday cages are reminiscent of the much mocked phrase "tin foil hat". It's now used to describe paranoid conspiracy theorists. The idea behind the lore is that there are people who believe malicious government agents are trying to send radio waves into their brains to either communicate a message or extract information out of their heads. If they wear a tin foil hat, the story goes, this blocks radio frequencies. A number of studies have proven there is some truth to this method (although they certainly didn't confirm the government was in fact beaming messages into people's heads). The problem is that tin foil hats aren't entirely closed, like a proper Faraday cage.

The EMP Commission report, among other expert advice, recommends placing some version of Faraday cages over as many important pieces of critical infrastructure as possible. It's not as outlandish as it seems. For example, the SCADAs discussed in a previous chapter should be enclosed in metal rather than sitting out in the open. Likewise, emergency generators should be covered up. Their job is to provide power in the event the main generators are knocked out. They certainly can't be exposed to the exact same risks the main generators face. If that were the case, then the emergency ones would just shut down alongside the ones they're supposed to replace.

SCADAs and generators might already be in Faraday cages by chance. Their cases or the room they're housed in may have the right metal or wire covering. But there's no way of telling until the issue is looked into. Hence the need for a plan. And it's important to understand the whole object needs to be covered. The doors need to be part of it too. A door-sized gap in the metal covering means it's not complete and won't have the same effect. It can't be a partial cover with a gaping hole, like those woefully helpless tin foil hats.

After the utilities are protected, another key

recommendation from the commission is to protect vehicles, particularly emergency and service ones. If the grid temporarily goes down but isn't too damaged or destroyed there's hope for a relatively short reboot. But there needs to be tools available to perform the repair job, which includes the vehicles needed to move the parts and tools and supplies around. As we've previously discussed, not every EMP wavelength hits everything the same. And the effects of solar storms and man-made attacks are slightly different. Depending on the wavelength, none, some or all of the vehicles will be impacted. This could be through being directly struck by an EMP. Or it could be through a wavelength getting into the vehicle via some other means. This is why two key recommendations are to keep emergency vehicles in proper Faraday cage garages, rather than in wooden ones or garages without doors that have a large, permanent opening. The other available measure is to insulate some of the main wires in the vehicle, to stop those wavelengths from getting at the interior of the vehicle and harming it that way.

The challenge once you start down this path is that it presents more questions than answers. What if the garage door is open at the time of the attack? Does that negate all the benefits? How tightly closed does a generator's Faraday cage shed need to be to prove resilient? What if there is a bit of wear and tear in the wire mesh? Does that mean you need to replace it? And what if you double up the wiring – does that create double the protection or somehow negate the protection entirely?

These questions actually point to one of the biggest recommendations from the EMP Commission report and gets to the heart of a lot of the legislative attempts made to tackle the issue. Before anyone calls for precise legislation requiring specific action, robust study is needed. Research departments and industries need to look at their operations and study how they'd be affected by an attack. Enumerate the challenges. List the different possible outcomes. Come up with mitigation strategies. Study. Plan. Test.

This is the big challenge for the entire subject. John F. Kennedy signed the atmospheric test ban treaty in 1963 right when scientists were just clueing into the importance of the subject. Learning was halted at a pivotal moment. Nobody is now calling for these bans to be lifted. They went a long way in guaranteeing the peace and stability most people in the West have enjoyed for decades. But because of the scarcity of real life examples – like the Hawaiian street light or Kazakhstan incidents – it's hard to know exactly how to protect society today because nobody knows for certain exactly how the objects and devices in our daily lives will respond to an attack.

Experts agree on general ideas about hardening the grid. And the scientific facts around something like Faraday cages are not disputed. Still, this lack of contemporary knowledge makes planning all that more difficult. It's possible the solution to this North American security risk is that we need extensive overhauls of many major facilities, across a number of sectors. Or it could just be that the few quick fixes noted above are all that's needed. Only study will answer these questions.

More study isn't needed, however, to confirm the basic, harrowing fact that if a solar storm or EMP hits power plants, transformers and transmission lines we'll be sorry that we did so little to protect them in advance.

When it comes to the power grid, the metal bags and the Faraday cages aren't the main event. No, the bigger fix is to add what are essentially large-scale surge protectors to the big ticket items in the electric system. Think about those stormy days when you're at home, lightning is flashing outside in the sky and the power goes out just for a moment and then everything powers back up again. The excess energy has been absorbed and thrown away, rather than fed into the system. Surge protectors defend against the risk of devices being overheated and damaged. They protect against electrical fires. This is why people buy surge protector cords for electronics like home computers and televisions. Just picture this on a much larger scale and that's what an extra high voltage transformer needs to handle an EMP attack.

The EMP Commission Report, and other expert testimonies, point to these giant surge protectors as the key to grid resiliency. They're the last step in a multi-pronged approach. The first step is to get monitors into place. If you can detect a major surge in power or an incoming pulse, you can take necessary steps to prevent or address it. Ideally, you can even unplug from the grid. If that doesn't work, the next line of defence is blocking devices that aim to deflect or otherwise mitigate the incoming surge. Lastly, the surge arrestors can absorb or redirect the power. Three pieces of technology to protect against an attack. All of them relatively inexpensive and easy to install. All of them made by North American firms. Yet few transformers and generators have these pieces. And no jurisdictions have standards or regulations for putting them in place.

This is what frustrates the likes of Boland and Pry so much. It's not like there are no tools available to address the problem. They're right there. Hidden in plain sight. An old fashioned personal desktop computer with a surge protector is more resilient to an EMP attack than the critical infrastructure that one of the most advanced societies in human civilization relies upon for survival.

# NORTHERN EXPOSURE

Energy knows no borders. Electrical currents don't carry passports. They go where the flow takes them. This certainly applies to an electromagnetic pulse moving through the atmosphere. But it also applies to energy on the ground, bouncing around the grid, running through the wires between the generators and transformer stations.

While there are electrical companies that are distinctly American or Canadian, and while the same applies to industry associations and regulatory agencies, the truth is that there is no distinction when it comes to energy itself. For the sake of electricity, the 49th parallel that separates much of these two countries is largely a blur. The North American grid largely consists of two major interconnected systems - an eastern and a western.

As the U.S. Department of Energy explains: "The Eastern Interconnection reaches from Central Canada Eastward to the Atlantic coast (excluding Québec), South to Florida and West to the foot of the Rockies (excluding most of Texas)... The Western Interconnection stretches from Western Canada South to Baja California in Mexico, reaching eastward over the Rockies to the Great Plains."

There are dozens of transmission points connecting Canada

and the U.S. Every Canadian province that's along the border is connected to the grid of at least one American state. In some cases, they're connected to more than one. Ontario, Canada's most populous province, is linked up to over a dozen states. This is nothing new either. Ontario first linked up with New York over a hundred years ago.

"This bilateral, bidirectional movement of electricity largely goes unnoticed by the public," a report by the Canadian Electricity Association from 2016 notes. "The home crowds cheering on the Vancouver Canucks might never contemplate that electricity generated in the U.S. could be illuminating the arena. Meanwhile, the car manufacturer in Michigan may be unaware that electricity from Canada is powering its assembly line."

Both of these systems are "electrically tied together", the Department of Energy explains. This is largely positive news for governments, industry and the consumer.
But not in every scenario. It means we share in each other's strengths, but we also share in catastrophe.

The massive 2003 northeastern blackout was the result of a number of issues, including aging infrastructure. But investigators found the catalyst was a tree branch touching a power line in Ohio. One branch. One line. And yet because of that, 45 million people in 8 states and 10 million people in Ontario went without power. Emergency measures were declared. Gridlock ensued. Many people were stranded. It was even responsible for a number of deaths, as has been explored in a previous chapter.

This is the innate challenge with the nature of such highly interconnected north-south power grids. It's also clear proof that EMP protection is something both countries must tackle with equal gusto. If Canada gets hit, the current will make its way through the grid to the United States. And vice versa.

There's a big problem though. Canada is lagging behind the

United States on the issue. Much further behind. Canada knows just a fraction of what most U.S. lawmakers, the media and the public know about the topic. And the Americans themselves hardly know much about the subject.

As we've detailed in previous chapters, there is some awareness and acknowledgement of the issue at the state and federal levels in the United States. The subject was classified for decades but the EMP Commission reports, starting with one delivered to Congress in 2004, brought the conversation out of the shadows and made it public. Even before then, there was chatter from various corners. Industry experts devised products for use by the military and concerned citizens. Researchers and scientists in engineering and physics departments around the country continued their study of the science behind EMP wavelengths and released reports and studies.

The fact the feds released money to harden the Cheyenne Mountain complex suggests that Barack Obama quietly acknowledged the subject and was at least willing to let that one initiative pass under his watch. (Although observers also worry that Admiral Gortney's retirement not long after publicly acknowledging EMP was no coincidence. Some believe he was forced out over this issue.) Protective legislation in America has appeared piecemeal. It's come in federal form courtesy of The Grid Act. It's arrived via state initiatives like those of Andrea Boland in Maine. This is the extent of America's efforts. At least it's something.

If only Canada were that far along. The only information of note just highlights how little is known and how nothing is being done. There has never been a single attempt to introduce legislation to protect the grid. The north is fully exposed.

In late 2016 and early 2017, I received responses to access to information requests that I had sent to the relevant government agencies that would be involved in protecting Canada from an EMP attack. These were National Defence,

Public Safety Canada, Royal Canadian Mounted Police (the RCMP is the federal police force, akin to the FBI) and the Canadian Security Intelligence Service (CSIS is similar to the CIA).

They were all asked to provide any reports, briefing notes, correspondence and so on that even so much as mentioned the EMP threat to Canada and / or the solutions being considered to combat this issue. While a couple of departments did hand over documents that mentioned geomagnetic storms in a peripheral way, the answer from them when it came to nuclear EMP weapons all was the same. Nothing. Absolutely nothing.

Public Safety Canada didn't even have it listed in documents tallying potential threats. CSIS had no documents even mentioning that this was a weapon that our enemies could possess. It's like the senior figures in these departments hadn't even heard about the subject before. In fact, the information officers certainly hadn't. In a number of cases, the simple request form I submitted left them perplexed. They had to call me to learn more about the issue so they could better search the documents.

Now even if there were reports that mentioned the issue but they were classified, the law in Canada governing access to information stipulates that they would still have to provide me with something. This would likely come in the form of highly redacted documents, dozens of pieces of blank paper or pages with marker blacking out every single line.

Curiously, before I had even made my request I had in my possession a report from 1987 that briefly discusses the issue. "Hazards for EDP Centres" was published by Emergency Preparedness Canada - the precursor to Public Safety Canada - in 1987 and authored by an employee of the evaluation and analysis department named Alain Drolet.

EDP sites, according to the report, are "electronic data processing" facilities. At the time the report was written storing data electronically was an emerging phenomenon. The brief five page report explains the various threats posed to these facilities. The list includes fires, floods, earthquakes, power outages, sabotage and war.

It's that last section where electromagnetic pulse makes its appearance. As a sub-section of the part on war, Drolet writes: "EMP (electromagnetic pulse) is also a possibility. A shielded room and the use of fibre optics for the lines that cross the border of the shield along with the filtration of power lines will greatly reduce the impact of EMP. Procedures can also be implemented to reduce the risk of EMP damage when sufficient warning is available. These procedures involve the physical disconnection of any long wires, such as telecommunication lines or power lines, from the equipment. One way to overcome some of these threats is to build a hardened site. This site will likely be underground, have its own air and water filtration (or recycling) plant, a water pumping system and an electric generator."

I didn't receive this report via brown envelope one day as some whistleblower secretly handed it over. It's posted on the department's online archives. And, the digital copy shows, it was immediately put into the government library system upon publication in 1987.

It's also telling that Drolet placed EMP in the 'war' category. During the Cold War, no less. Did they know something we didn't? Was the Canadian government secretly fearing an EMP attack from the Soviets? These days, the U.S. electrical industry reports that mention EMP refer to the issue almost exclusively along the lines of naturally occurring EMP, or geomagnetic disturbance or solar storm. They go out of their way to avoid referencing the fact that it is also a weapon of war. Yet a Canadian government report casually notes in a routine report about protecting data centres that it is in fact part of war. This is an unsolved mystery.

Bizarrely, the department didn't provide Drolet's report to

me when I submitted my request. They certainly should have, seeing as I requested any reports mentioning the topic going back to the 1960s. Based on the rule that the simplest explanation is likely the correct one, I chalk this slip up to the challenges of file management. It's not like they would have opened up every single report and flipped through every page to look for a mention of the issue.

It's still curious that there is at least this one report casually mentioning the issue as if it's routine knowledge known by most government employees, yet there is no major report addressing the topic and its dangers. Or if there is, it's yet to be released.

Senator Daniel Lang has had the same experience. He's the closest thing Canada's got to an Andrea Boland. And, like her, he stands alone. 2016 was the year Canada started, ever so slightly, a conversation about EMP. It was all because of Lang's efforts.

The Conservative Senator is in some respects an uncharacteristic individual to champion the issue. He represents Yukon – a northern territory bordering Alaska that only had its first elections in 1978 and wasn't properly formalized as its own government until 2003. Once I called Lang's office to speak with him and was told he wasn't in his Parliament Hill office in Ottawa, Ontario, but back home in the Yukon. I said I was fine with him calling me back from wherever he was in Yukon. The only problem with that, his staff member said, was that Lang was out in the deep woods on a hunting expedition and no one knew when he was expected back.

All of this is to say that Yukoners are much more self-reliant individuals than the average North American. Living off of the grid isn't that much of a stretch for them. They're probably the people who could best survive a catastrophic EMP attack. Yet it's a politician from Yukon who is leading a one man charge to protect the nation.

In April 2016, Lang convened a hearing on the issue in his capacity as chairman of the Senate standing committee on national security and defence. The expert panel that came from the United States included Peter Pry. It was the first government forum on the subject in Canada.

Then, weeks later, Lang formally requested in the Senate that the governing Liberal party move to declassify the subject entirely. "The government owes it to public, policy planners, industry and first responders at local, regional and national levels to have access to all information on the threat and be afforded the opportunity to take mitigation measures," Lang said at the time.

Eight months after Lang submitted the request, he finally received an answer. Or perhaps non-answer is the better way to describe what he received. "When will electromagnetic pulse information be declassified?" is the key question, written in bold, that the document Lang received strives to answer. But immediately below this question is written: "The Department does not do any work related to nuclear EMP." Natural Resources Canada was the department the Senate clearly assigned to answer the question. They won't be declassifying the subject because there's nothing to declassify.

It was curious to see that the official response came from Natural Resources Canada, which is not one of the federal departments that typically engages in national security threat assessments. It became clear though why this department was chosen to offer response once I discovered who in the department helped pen the response.

While the vast majority of Canada's 258,979 federal government employees (according to 2016 numbers) remain unaware of the subject of EMP, there is at least one person who specializes in the topic. David Boteler is a government scientist who heads up the government's geomagnetic laboratory in Ottawa. Several of his research papers on geomagnetic disturbances have informed the writing of this book. Not only has Boteler partnered with scientists from around the world to analyze the effects of solar flares on earth and power grids, but he's worked with Hydro-Quebec experts to help devise mitigation strategies for the next time a storm hits. The only problem is the discussion of geomagnetic storms often happens completely disconnected from nuclear EMP in official circles. This is largely due to the fact that, as previously discussed, the underlying waveforms behind the two events are different. The good news is there are at least a handful of people in the government concerned about a repeat of the Carrington Event. The bad news is there doesn't seem to be anyone working on nuclear EMP.

Although the plot thickened at another Senate committee hearing that happened a couple weeks after Lang made his request. A Public Safety Canada bureaucrat spilled the beans on the subject. The news still wasn't good though. Colleen Merchant, director general of Canada's National Cyber Security Directorate, admitted the country doesn't have a coordinated plan to deal with an attack.

"In terms of the EMP," Merchant explained, "if it were to have a significant effect on the Canadian infrastructure, it would be very important for us to have a plan at hand to respond and understand roles and responsibilities of the federal government, provinces and territories, first responders and the critical infrastructure owners and operators." Just like Alain Drolet almost 30 years prior, Merchant clearly knew about it. But how much does she know? Is she unique in this capacity? Do many of her colleagues know?

There is optimism that Prime Minister Justin Trudeau is aware of the subject and willing to move forward on it. Despite his well-known boxing skills and alpha male good looks, Trudeau is in many respects something of a geek. He dresses up in Star Wars outfits to take his kids out on Halloween and famously gushed enthusiastically to the press about his amateur passion for quantum computing. He appears well poised to have an interest in electromagnetic pulse.

One innovation of Trudeau's government is that they publicly posted the ministerial mandate letters given to each cabinet minister. This is the formal document bearing the prime minister's signature instructing members of cabinet what their legislative goals are for the upcoming session.

The fourth of 12 priorities in the mandate letter to Ralph Goodale, the minister for public safety and emergency preparedness, instructs him to "lead a review of existing measures to protect Canadians and our critical infrastructure from cyber-threats." Whether or not EMP is a cyber threat is an issue of semantics. Sometimes it falls under this category, other times it does not. It all depends. Goodale's office has declined to respond to questions about whether this priority will include looking into the EMP issue. It could all be a coincidence. Or Justin Trudeau could very much be aware of the EMP issue and intent on addressing it. So far, it's unknown.

In September of that year another Canadian politician finally spoke up on the issue. Tony Clement, a Member of Parliament and former cabinet minister from the previous government, was running for leader of the Conservative Party of Canada following the departure of Stephen Harper. His national security platform included enhanced funding for critical infrastructure resilience that specifically included EMP grid protection. However before the end of the year Clement withdrew from the leadership race. No other candidate has adopted this idea.

It is also unclear, at the time of this writing, how the United States under the administration of President Donald Trump will engage with this issue, if indeed it acknowledges it at all. Former CIA director James Woolsey was originally an advisor on Trump's election campaign and subsequent transition team. Woolsey has been one of the most vocal advocates of EMP awareness, co-authoring op-eds with Peter Pry. However, Woolsey resigned from the transition team prior to inauguration day. Early indications of Trump's financial plans reveal cuts to the Department of Energy that "would roll back funding for nuclear physics and advanced scientific computing research to 2008 levels," according to a January 2017 report in The Hill Times.

Until Trudeau or one of his ministers speaks up on the issue, or until Senator Lang's declassification request gets traction, or until research efforts yield further fruit, the issue in Canada is at a standstill.

There is one saving grace though. A full quarter of Canada's energy is hydroelectric. Water flows through dams, spins turbines and generates energy, which the power company stores and ships out. An EMP attack may damage the parts at these plants, but it can't of course stop water from flowing. It's a natural power source that's not going away. The United States, according to a 2016 report from energy giant BP, only gets 2% of its energy needs from hydro. Canada's hydroelectric advantage is still no excuse for inaction and ignorance. And right now that's exactly where Canada finds itself.

# THE DESCENT OF THE BLACK SWAN

Tucked away in the cabinet above our refrigerator we have a package of pills in a white envelope. They're called KI pills. Those are the periodic table letters that spell out potassium iodide.

We have them in the event of an emergency occurring at one of the nuclear power plants close to our home in Toronto. The goal is to take them around the time of exposure to radiation to reduce the chances of developing cancer. There are twenty of them, enough for my spouse and I and our two children. They don't expire until 2027.

We didn't seek these out. We weren't reading some junk science website or fringe medicine report about it. We didn't order them from a basement blogger hawking emergency preparedness material. We got them from the municipal government. Or rather, they sent them to us after we filled in a simple online form. We're not particularly worried about a nuclear catastrophe happening in this region. But it made sense to take the government up on the offer they were promoting at the time. Better safe than sorry.

In fact, the Canadian Nuclear Safety Commission, a federal agency, mandates that anyone who lives within 10 kilometres of a nuclear station receive these pills. And those

given how few reputable resources the general public has to understand the topic. But the blurring of fact and fiction got so bad in the eyes of the U.S. military that they commissioned popular science personality Bill Nye the Science Guy to make a corporate video for internal use only explaining the differences between EMP in pop culture and the truth about EMP. The video is not available for public use. Nye would not comment on it when his representatives were contacted with an interview request.

Despite the fact the issue is real and has even leaked into popular culture, inaction remains a problem. This is a danger to national security. As North America does nothing, the black swan descends.

This phrase comes from hedge fund manager turned academic Nassim Nicholas Taleb's influential 2007 book *The Black Swan: The Impact of the Highly Improbable.* The phenomenon, Taleb writes in the book's preface, has three principal characteristics: "First, it is an *outlier,* as it lies outside the realm of regular expectations, because nothing in the past can convincingly point to its possibility. Second, it carries an extreme impact. Third, in spite of its outlier status, human nature makes us concoct explanations for its occurrence *after* the fact, making it explainable and predictable."

Examples of recent black swans range from the staggering success of tech company Google and how it's changed the way we find, store, receive and even think about information, to the tragic 9/11 attacks, the reverberations of which will be felt for decades to come.

"Why do we not acknowledge the phenomenon of black swans until after they occur?" the book's jacket description asks. "Part of the answer, according to Taleb, is that humans are hardwired to learn specifics when they should be focused on generalities. We concentrate on things we already know and time and time again fail to take into consideration what we don't know. We are, therefore, unable to truly estimate opportunities, too vulnerable to the impulse to simplify, narrate, and categorize, and not open enough to rewarding those who can imagine the "impossible."''

Along the same lines, Peter Pry describes how we haven't grasped the importance of EMP as a "failure of strategic imagination." The whole point of a surprise attack is that it is a surprise. The goal of a game changer is to change the game. Old rules don't apply. Past performance is not an indicator of future outcomes, as the phrase found at the bottom of many investment prospectuses explains. Many people intellectually understand this rule. Few people act accordingly.

This idea matters now more than ever. The likelihood of black swans is increasing. And their impacts are worse. Taleb's preface continues: "A small number of Black Swans explain almost everything in our world, from the success of ideas and religions, to the dynamics of historical events, to elements of our own personal lives. Ever since we left the Pleistocene, some ten millennia ago, the effect of these Black Swans has been increasing. It started accelerating during the industrial revolution, as the world started getting more complicated, while ordinary events, the ones we study and discuss and try to predict from reading the newspapers, have become increasingly inconsequential."

By referencing the increasing complexity of the world, Taleb is unintentionally referring to the electronic civilization that has made our lives so fragile and vulnerable. The more connected we are and the more advanced we are, the larger the black swan that will swoop down upon us. An electromagnetic pulse attack is the mother lode of black swans. When it happens, it will change everything. And I write "when it happens" and not "if it happens" on purpose. For one reason only. We are doing nothing about it despite both the problem and the solution being hidden in plain sight. It seems only inevitable that, until grid protection becomes an issue, North America's enemies will keep this weaponry as an option in their strategic plans.

It's in Russia's playbook. They've used it before over Kazakhstan. Iran and North Korea know about it, if they don't already have it. It's only a matter of time before terrorist groups and organized crime attempt to deploy an EMP device.

The worst case scenario EMP attack will come with no warning and will only happen once. After the fact, society as we know it will be destroyed. It will be too late by then to do anything about it. The time to act is now.

This is a difficult issue for the public and policy makers to grasp. Normally, we only act on issues after they've started to build in the public consciousness and we've seen some of their minor negative impacts. When it comes to, say, a rise in certain crimes or a specific form of cancer, we gradually hear about cases impacting the lives of people in our communities. We learn through experience just how bad these things are and it's this process of learning that precipitates public awareness and action.

It's possible that a small scale EMP attack will happen in a major city. It may only take out a handful of transformer stations, blacking out a community of several million for days or weeks. There will be hardships and a handful of deaths among the most vulnerable, but the city will recover. This will be the spur to action North America and even the rest of the world needs for protection. However there is no guarantee there will be such a learning curve. The first attempt could be the worst. And therefore the last.

The stories explored in this book don't tell us an attack is imminent. But they show the pieces are coming together. The mysterious and highly professional Metcalf transformer station attack. The North Korean Kwangmyŏngsŏng-3 that passes over the centre of continental North America. The Carrington Event that damaged electronics around the world. The Hawaiian streetlight incident that first alerted the American government as to just how bad the problem could get. The Maroochy water incident in which a single disgruntled employee took over a utility company innocent civilians relied upon. The arming of the Chong Chon Gang – the vessel traveling between North Korea and Cuba that alarmingly had parts used in the building of EMP weapons. Those military strategies form the likes of General Vladimir Slipchencko in Russian textbooks. The US military being concerned enough to spend hundreds of millions on hardening Cheyenne Mountain. The examples continue.

We don't want to wait to act until the picture is complete. You either work to protect yourself against an attack before it happens or you do nothing and suffer the consequences.

# NOTES AND BIBLIOGRAPHY

This book is intended for the general reader and is meant as a primer on the subject of EMP and its potential impacts on North America. While there is no other title published to my knowledge in the English language that puts together all the threads explored in this book, there are excellent papers and reports available that do explore in depth different elements addressed here.

For those readers concerned about this issue and wishing to raise awareness of it, consider forwarding a copy of this book to a politician or influential member of your community.

The first stop a reader looking for more information should make is to the EMP Commission to download their comprehensive reports. They are available at www.empcommission.org.

The next stop should be to several books by Peter Vincent Pry available at Amazon.com, such as 'Blackout Wars". These are essentially a continuation of Pry's work at the commission.

The science behind geomagnetic storms and EMP is studied in detail in a number of academic and government publications that are either directly quoted in this book or have greatly informed it. They include:

"The super storms of August/September 1859 and their effects on the telegraph system", by D.H. Boteler, published in Advances in Space Research 38 (2006), pp.159 – 172

"A Colorful Blackout: The Havoc Caused by Auroral Electrojet Generated Magnetic Field Variations in 1989," by Sebastian Guillon, Patrick Toner, Louis Gibson and David Boteler, published in IEEE Power & Energy Magazine,

November / December 2016

"Geomagnetic Effects on Power Systems," by D.H. Boteler, L. Trichtchenko and R. Pirjola, co-published by the Geomagnetic Laboratory of Natural Resources Canada and the Finnish Meteorological Institute, Helsinki, Finland.

"Justification and Verification of High-Altitude EMP Theory Part 1", by Conrad L. Longmire, prepared for the Lawrence Livermore National Laboratory by the Mission Research Corporation in June 1986

"The Uncertain Consequences of Nuclear Weapons Use," by Michael Frankel, James Souras, George Ullrich; Johns Hopkins Applied Physics Laboratory, 2015

"Did High-Altitude EMP Cause the Hawaiian Streetlight Incident?" by Charles N. Vittitoe, Sandia National Laboratories, 1989

"High Altitude Electromagnetic Pulse (HEMP) and High Power Microwave (HPM) Devices: Threat Assessments", a Congressional Research Service Report by Clay Wilson, published in Terrorism: Commentary on Security Documents volume 119, Catastrophic Possibilities Threatening U.S. Security, by Oxford University Press, January 2012

"Project Officer's Interim Report: Starfish Prime", prepared by Ray L. Loadabrand and Lambert T. Dolphin Jr, August 1962, prepared for Field Command Defense Atomic Support Energy at Sandia Base, Albuquerque, New Mexico

"The Early-Time (E1) High-Altitude Electromagnetic Pulse (HEMP) and Its Impact on the U.S. Power Grid", by Edward Savage, James Gilber William Radasky. Produced by the Metatech Corporation, January 2010

"Consideration for a Power Transformer Emergency Spare Strategy for the Electric Utility Industry" was prepared by the Electric Power Research Institute for the US Department of Homeland Security Science and Technology Director, September 2014

"USSR Nuclear EMP Upper Atmosphere Kazakhstan Test 184" is a report by the Electric Infrastructure Security Council.

"Initial Economic Assessment of Electromagnetic Pulse (EMP) Impact upon the Baltimore-Washington-Richmond Region" produced by The Sage Policy Group, September 2007

The Arms Control Association (www.armscontrol.org) is an authoritative source on the history of nuclear test launches.

The section on the ENIAC was greatly informed by the stellar research in the 2016 book "ENIAC in Action: Making and Remaking the Modern Computer" by Thomas Haigh, Mark Priestley and Crispin Rope.

Science writer Stuart Clark compiles fascinating information on The Carrington Event in "The Sun Kings: The Unexpected Tragedy of Richard Carrington and the Tale of How Modern Astronomy Began".

Journalist Rebecca Smith, writing in the Wall Street Journal, has written detailed accounts of the Metcalf transformer station attack, beginning in February 2014. Her reporting was an invaluable resource.

To further appreciate how fragile the electronic civilization is you don't need to look any further than government reports, such as the U.S. Department of Energy's "Final Report on the August 14, 2003 Blackout" and their "Large Power Transformers and the U.S. Electric Grid". Both are available on their website.

"One Second After" is a powerful novel by American history professor and popular writer William R. Forstchen. It tells the story of what happens following an atmospheric EMP attack from the eyes of a family in small-town America.

I have consulted extensive files and research on this subject but am always eager to unearth more studies and hear from experts, particularly those working within governments. Hopefully there will be further revelations on this subject, and once those occur I expect to release a second edition of this book.

I can be reached at anthonyfureywriter@gmail.com.

# ACKNOWLEDGEMENTS

It's a great privilege to work as a syndicated newspaper columnist. The most rewarding part is receiving correspondence and feedback from readers of all walks of life. The significant and positive feedback to stories I wrote on EMP in January 2016 is what inspired me to continue my research and compile this book. I'd like to first of all thank those readers for encouraging me and supporting my work.

My sincere thanks goes out to those who either read drafts, offered advice or gave much appreciated encouragement, including Patrick Furey, Brenda McMillan, Doug May, Candice Malcolm, Kasra Nejatian, Sergei Voitchenko, Michael Levine, Naresh Raghubeer, Jane Bigley, David Harris and others.

Special thanks goes to my wife, Leslie, who supported me throughout the process.

# ABOUT THE AUTHOR

Anthony Furey is a national columnist and editor for Postmedia, Canada's largest chain of newspapers and news sites. He was previously comment editor of the Ottawa Sun as well as a national correspondent based out of the nation's capital.

He's also written for TIME, NY Daily News, Human Events and more. He's been a guest on Fox News Channel, BBC and various other programs.

A familiar voice on talk radio, Furey currently hosts a morning talk show on SiriusXM that broadcasts across North America.

Furey studied philosophy and history at the University of Toronto.

© 2017 – Anthony Furey